Little Acorn Books

About Early Math

EARLY MATH SKILLS PRACTICE FUN

by Marilynn G. Barr

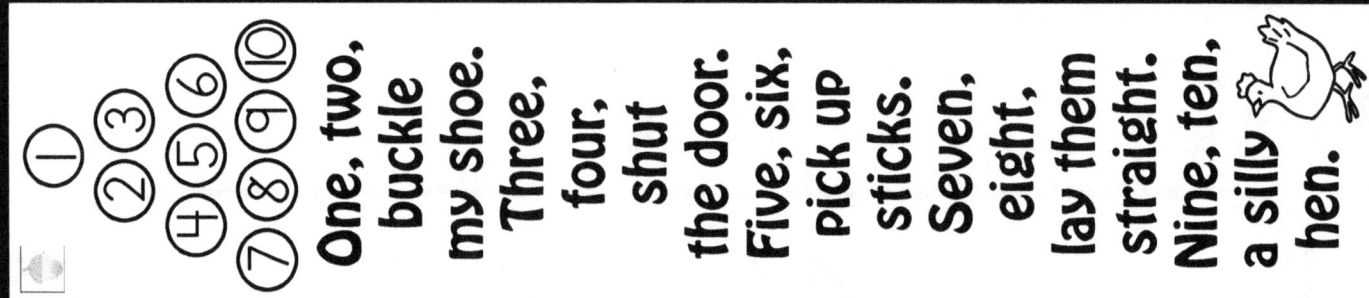

LAB201410P
ABOUT EARLY MATH
Early Math Skills Practice Fun
Preschool — Grade 1
(*Skills Focus: readiness skills, counting, writing numerals and number words; counting and matching numerals, number words and number sets; telling time, number recognition, early addition and subtraction, measuring, skills achievement awards*)

by Marilynn G. Barr

Published by: Little Acorn Books™
Originally published by: Monday Morning Books, Inc.

Entire contents copyright © 2014 Little Acorn Books™

Little Acorn Books
PO Box 8787
Greensboro, NC 27419-0787

Promoting Early Skills for a Lifetime™

Little Acorn Books™
is an imprint of Little Acorn Associates, Inc.

http://www.littleacornbooks.com

Permission is hereby granted to reproduce student materials in this book for non-commercial individual or classroom use. *School-wide or system-wide use is expressly prohibited.

ISBN 978-1-937257-56-9
Printed in the United States of America

About Early Math

Contents

Introduction 4
Number Charts
One Dancing Elephant............................. 6
Two Meowing Cats................................... 7
Three Quiet Bears 8
Four Swimming Ducks 9
Five Funny Monkeys 10
Six Muddy Pigs 11
Seven Singing Koalas 12
Eight Silly Snakes 13
Nine Clapping Butterflies 14
Ten Hopping Bunnies............................. 15
Eleven Smiling Frogs 16
Twelve Tiny Turtles 17
Match Boards
Elephant Match Board 18
Cat Match Board................................... 19
Bear Match Board 20
Duck Match Board 21
Monkey Match Board 22
Pig Match Board 23
Koala Match Board................................ 24
Snake Match Board 25
Butterfly Match Board............................ 26
Rabbit Match Board............................... 27
Frog Match Board 28
Turtle Match Board................................ 29

Number Set Cards
Peanut Cards 30
Cookie Cards 31
Button Cards 32
Drum Cards 33
Bell Cards 34
Flower Cards 35
Key Cards... 36
Star Cards 37
Bow Cards.. 38
Carrot Cards 39
Balloon Cards 40
Heart Cards 41
Numeral Cards 42
Number Word Cards.............................. 43
Add or Take Away 44
Game Board 45
Addition Problem Cards 47
Subtraction Problem Cards..................... 50
Math Problem Boards 53
Number of the Day Badges 54
Number Charts
Numbers Chart 55
Number Words Chart............................. 56
Awards
 Awards Booklet.................................. 57
 Awards 59
Take-Home Notes 60
Clock Form 61
Counters 62
Animal Inches 64

About Early Math

Introduction

One Dancing Elephant, Five Funny Monkeys, Twelve Tiny Turtles and nine more delightful creatures provide plenty of early math skills practice for little ones. *ShortCuts for Early Math* includes number charts, match boards, a clock form, and number charts for numbers 1-12. Number set cards are designed for use with both match boards and math problem boards. Add or Take Away, a trail game, provides early addition and subtraction skills practice for numbers 0-10. Counters are also included for children to practice counting to 100. *About Early Math* is one of four-book series which includes: *About Early Writing, About Early Colors & Shapes,* and *About Early Reading.*

Number Charts

Reinforce hands-on counting and number recognition skills with the twelve number charts on pages 6-17. Reproduce number charts for children to color and cut out. Prepare a work station with a basket of large buttons for children to glue on number charts. Mount finished charts on construction paper for display.

Reproduce a set of number charts for each child to color. Have children decorate construction paper covers. Then help each child stack and staple number charts in numerical order to form a book.

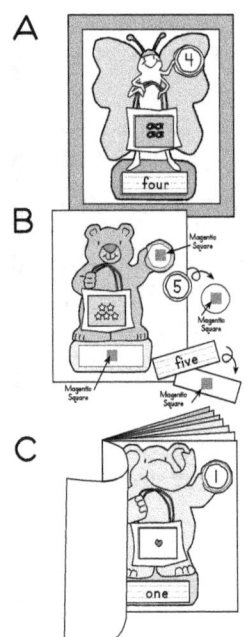

Match Boards

For a sorting, matching, and visual discrimination skill building activity, reproduce match boards (pages 18-29), number set cards (30-41), numeral cards (page 42), and number word cards (page 43) for children to color and cut out. Have children sort, match, then glue one matching number set, numeral, and number word card on each match board. Mount finished boards on construction paper for display. *Diagram A*

Reproduce, color, and cut out the match boards. Glue one number set card (1-12) on each match board. Laminate, punch two holes, lace, and tie yarn through the holes in each board to form a book. Reproduce, color, cut out, and laminate a set of numeral and number word cards. Attach a magnetic square on the back of each card and blank spaces on each match board. Store the loose cards in a resealable plastic bag. Children sort and attach the matching numeral and number word cards on each match board. *Diagram B*

Reproduce a complete set of match boards and one set of number set, numeral, and number word cards for each child. Have children color, cut out, and glue the card cutouts on match boards. Then help each child stack and staple assembled match board pages in numerical order to form a book. *Diagram C*

Add or Take Away

Children practice adding and subtracting as they play Add or Take Away (pages 44-52).

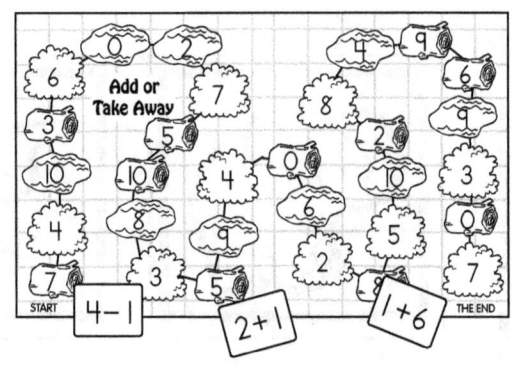

To play, each of two to four players chooses one of the twelve animal pawns. One player shuffles and places the deck of cards, face down, on the table. Each player, in turn, draws a card, solves the addition or subtraction problem on the card, and moves his or her pawn to the next matching numeral space on the board. Drawn cards are placed, face down, in a discard pile. Play continues until each player reaches The End. When all the cards have been drawn, reshuffle the discard pile and continue playing.

Math Problem Boards

Create several sets of Math Problem Boards (page 53) for children to practice adding and subtracting. Program math problem boards with number set cards, then reproduce, color, laminate, and cut out the boards. Reproduce, color, cut out, and laminate number set answer cards to match the programmed math problem boards. Children can work individually or in two-person teams to solve the problem boards. Store boards and answer cards in decorated manila envelopes.

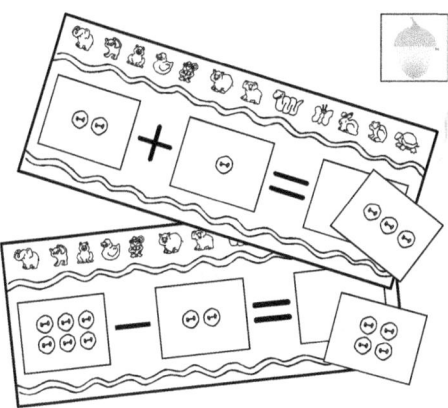

Number of the Day Badges

Program and reproduce oak tag badges (page 54) with the "Number of the Day" for each child. Have children color and cut out badges. Use cellophane tape to attach safety pins to the backs of badges for children to wear. Option: Provide each child with a folder. Have children glue badges to the fronts of folders. Write each child's name on the front of his or her folder. Children can use these folders to store completed or unfinished work related to the "Number of the Day" on the cover.

Number Charts

Reproduce, color, cut out, and laminate a set of charts (pages 55-56) to post on a display board or in a math skills practice center. Set up a table with a bowl of cereal Os for a hands-on math skills practice activity. Children count and place the matching number of Os on each space on a chart. Number Charts can also be used with number set cards. Reproduce number set cards for children to match on the charts.

Awards

Children love to receive awards and keep track of their own achievements. Reproduce award booklets (pages 57-58) for children to color and cut out. Help each child assemble and staple his or her booklet. Store all star stickers in a basket. Give children star stickers to glue in booklets as they master each listed skill.

Be prepared to reward children for math skills achievement. Reproduce, color, and cut out a supply of awards (page 59). Store awards in decorated envelopes or folders.

Take-Home Notes

Send home notes to keep parents informed about current math skills practice. Reproduce and cut apart a supply of bright-colored paper notes (page 60). Store notes in a decorated manila envelope or folder.

Clock Form

Prepare children for telling time with the clock form on page 61.

Counters

Children can practice counting to 100 with the animal counters found on pages 62-63.

Animal Inches

Introduce children to measuring with the Animal Inches ruler on page 64.

One Dancing Elephant

One Dancing Elephant

1

Two Meowing Cats

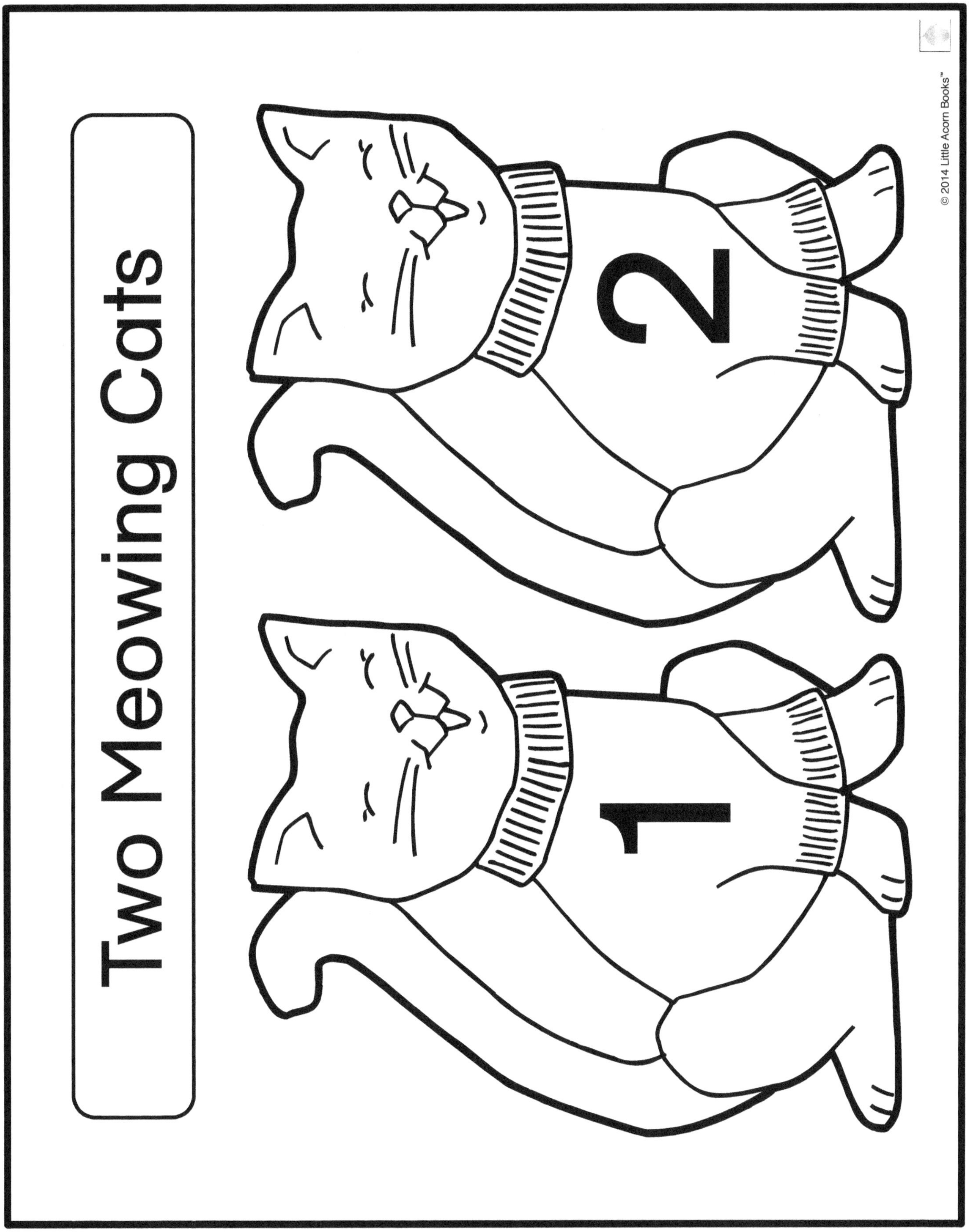

Three Quiet Bears

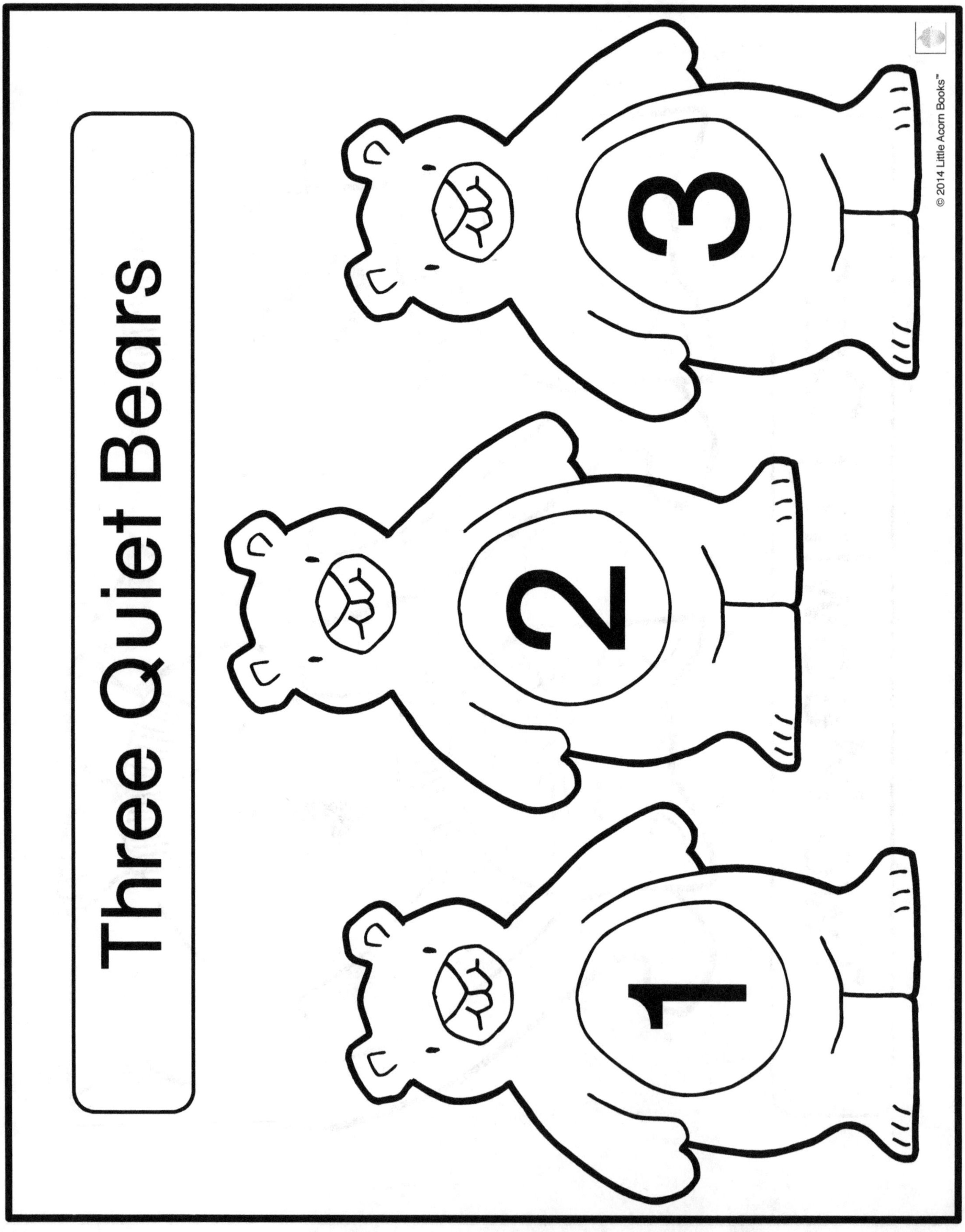

Three Quiet Bears

3

2

1

Four Swimming Ducks

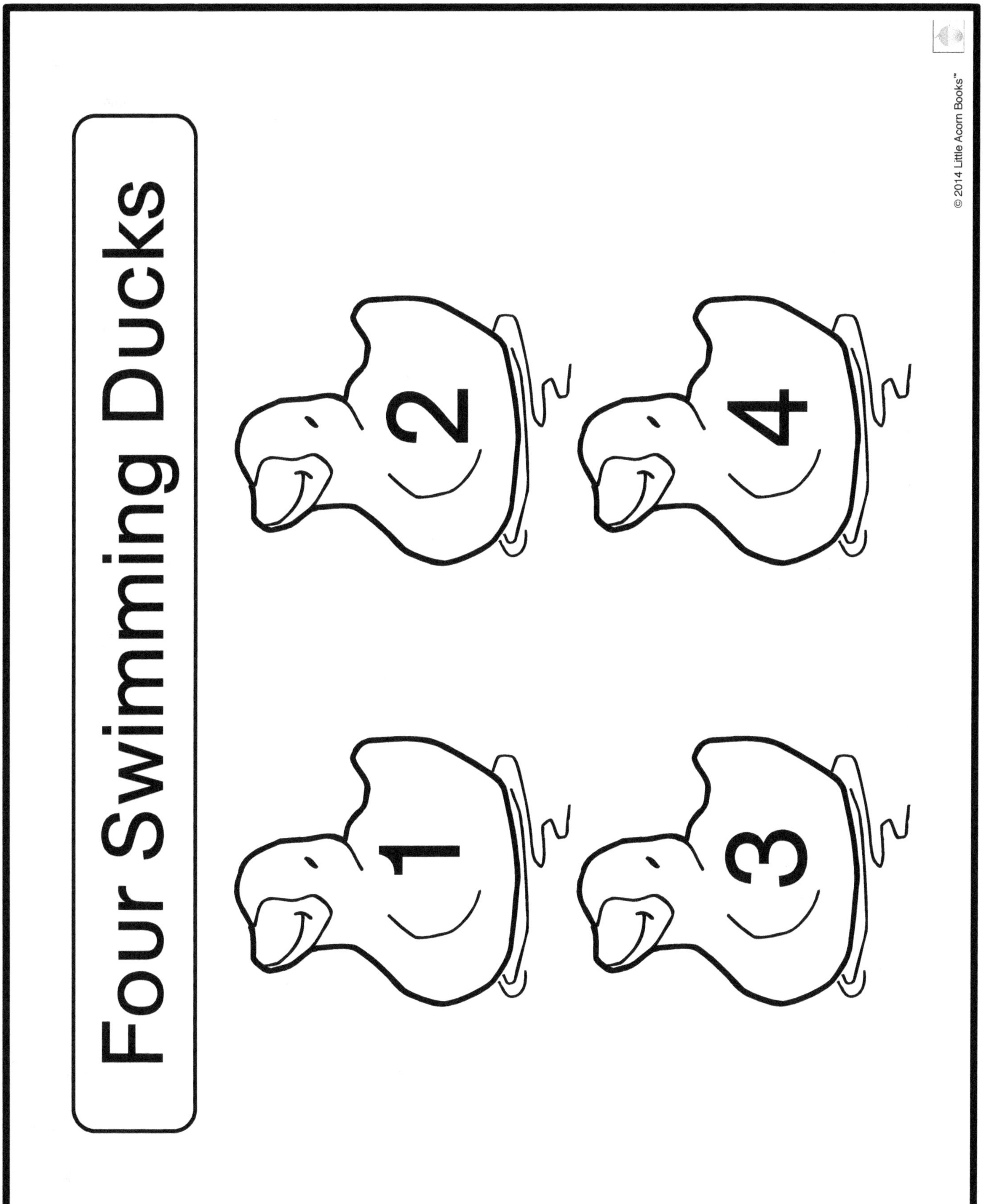

Four Swimming Ducks

Five Funny Monkeys

Six Muddy Pigs

Six Muddy Pigs

3   6

2   5

1   4

Seven Singing Koalas

Eight Silly Snakes

Nine Clapping Butterflies

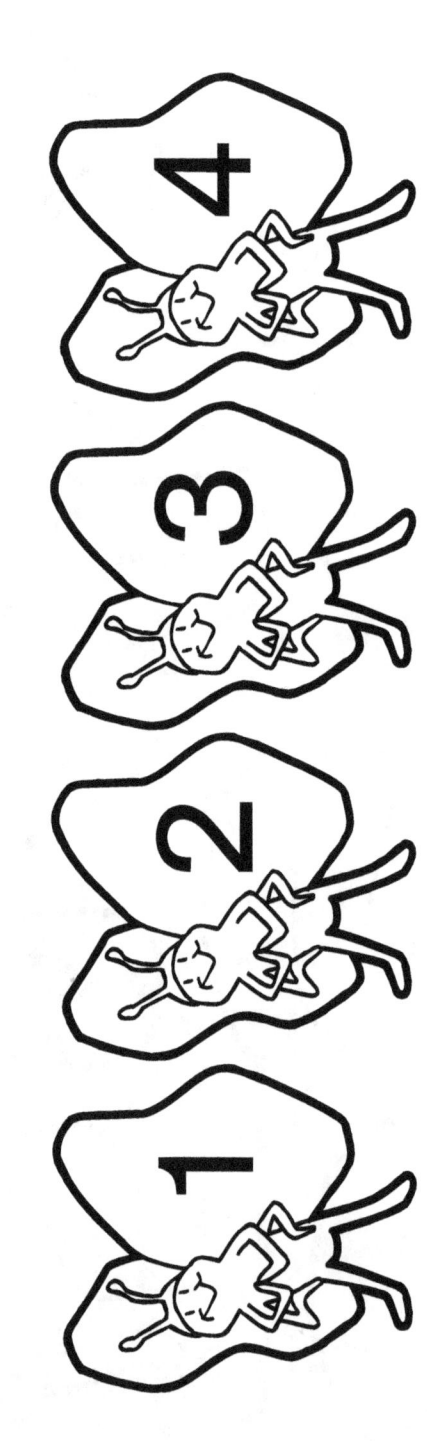
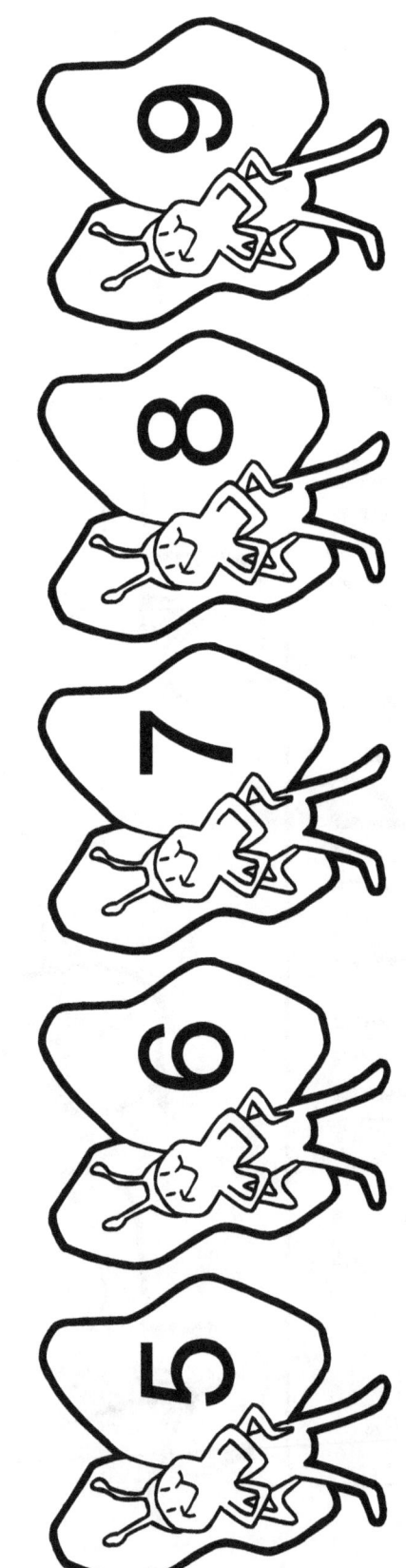

Ten Hopping Bunnies

Eleven Smiling Frogs

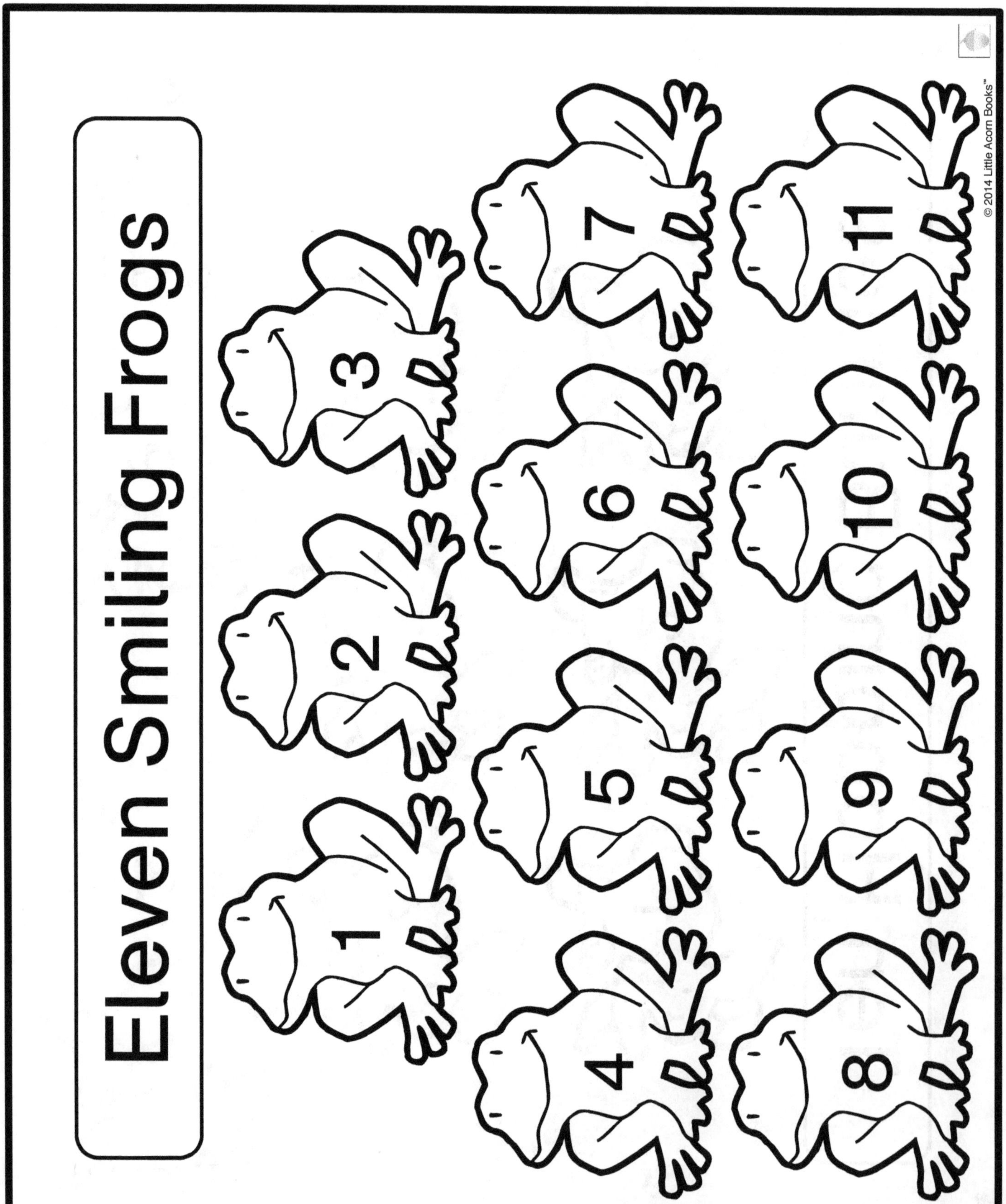

LAB201410P • About Early Math • 978-1-937257-56-9 • © 2014 Little Acorn Books™

Twelve Tiny Turtles

Elephant Match Board

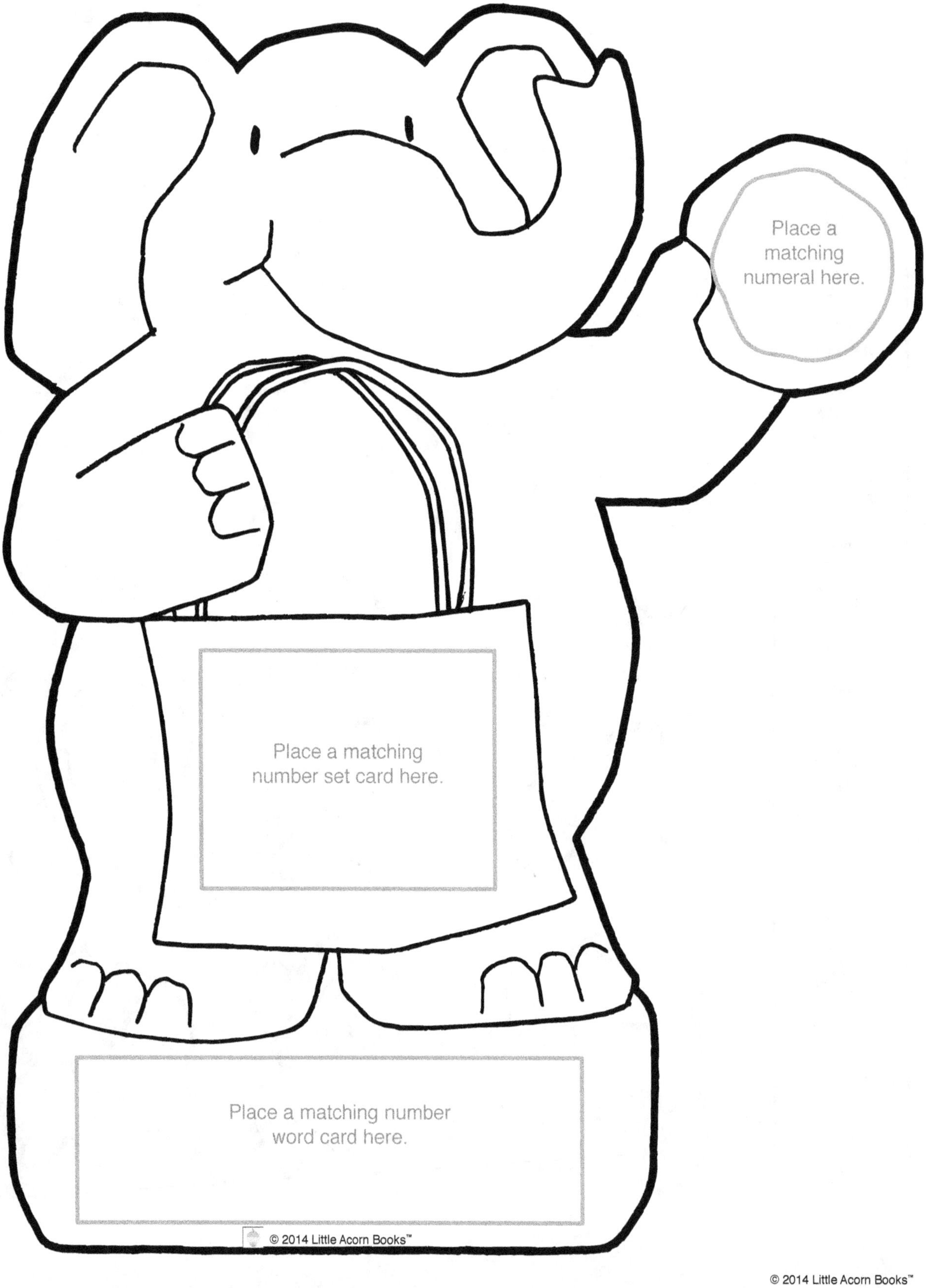
Place a matching numeral here.
Place a matching number set card here.
Place a matching number word card here.
© 2014 Little Acorn Books™

Cat Match Board

Place a matching numeral here.

Place a matching number set card here.

Place a matching number word card here.

© 2014 Little Acorn Books™

Bear Match Board

Place a matching numeral here.

Place a matching number set card here.

Place a matching number word card here.

© 2014 Little Acorn Books™

Duck Match Board

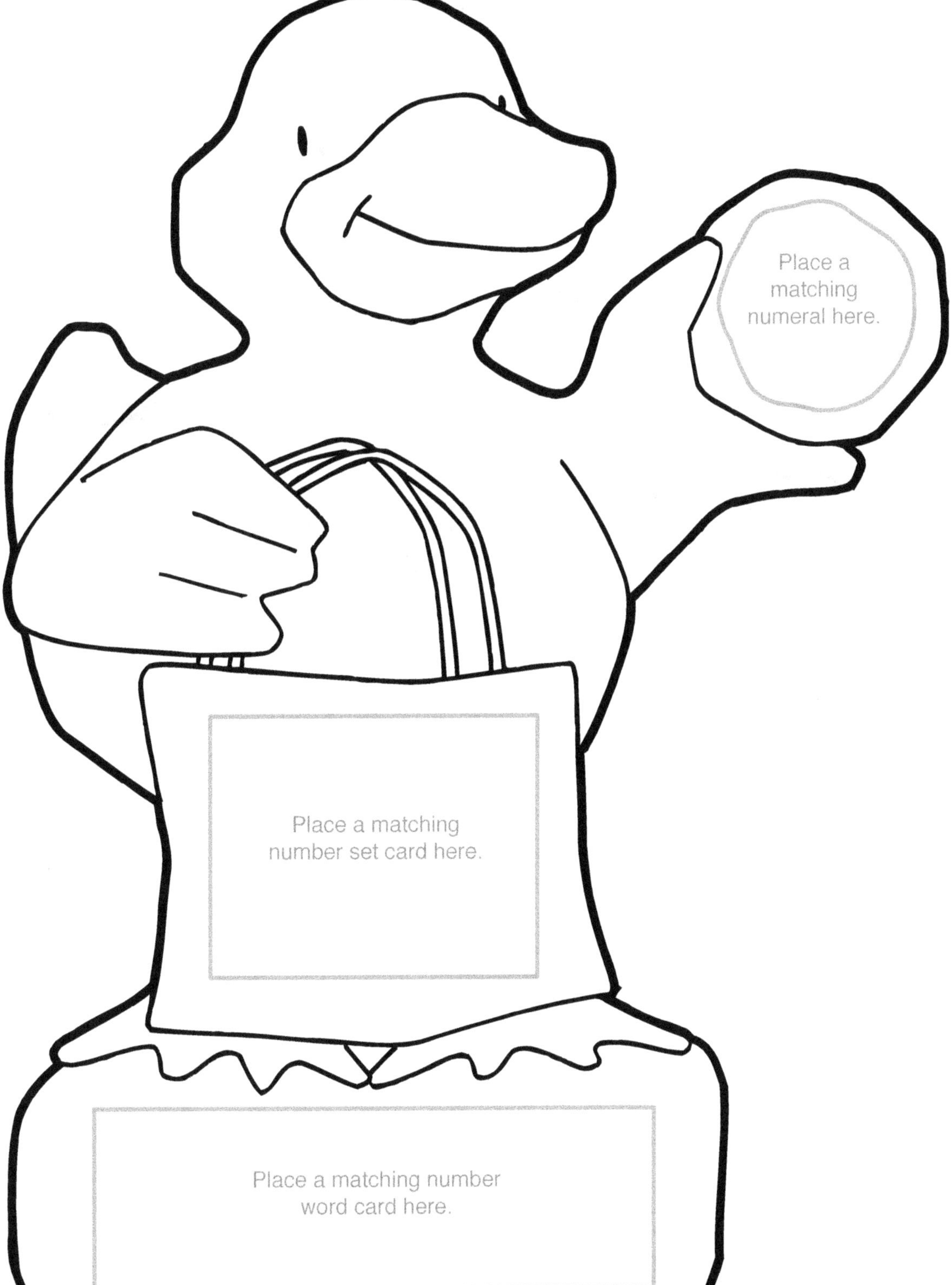
Place a matching numeral here.
Place a matching number set card here.
Place a matching number word card here.
© 2014 Little Acorn Books™

Monkey Match Board

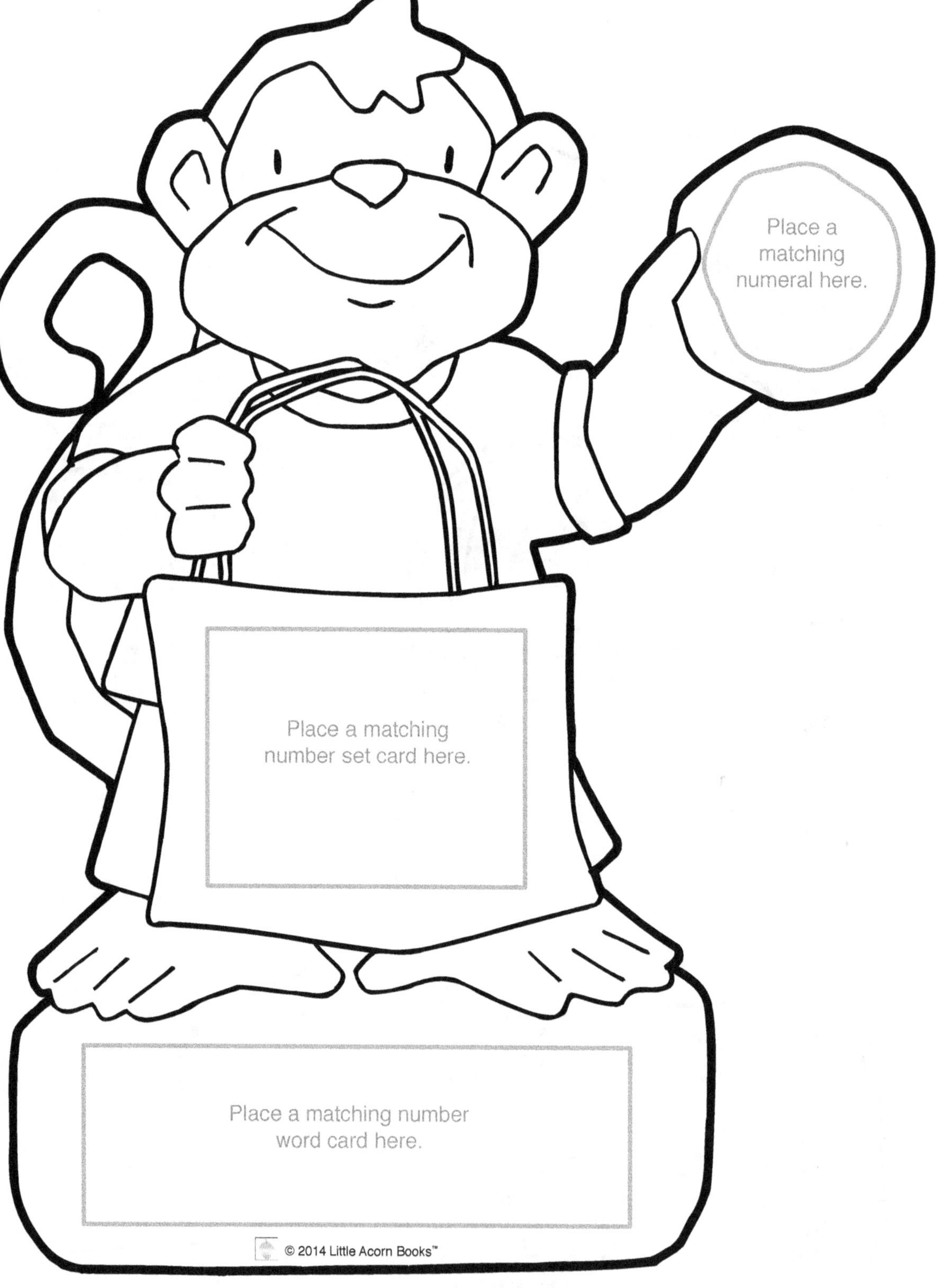
Place a matching numeral here.
Place a matching number set card here.
Place a matching number word card here.
© 2014 Little Acorn Books™

Pig Match Board

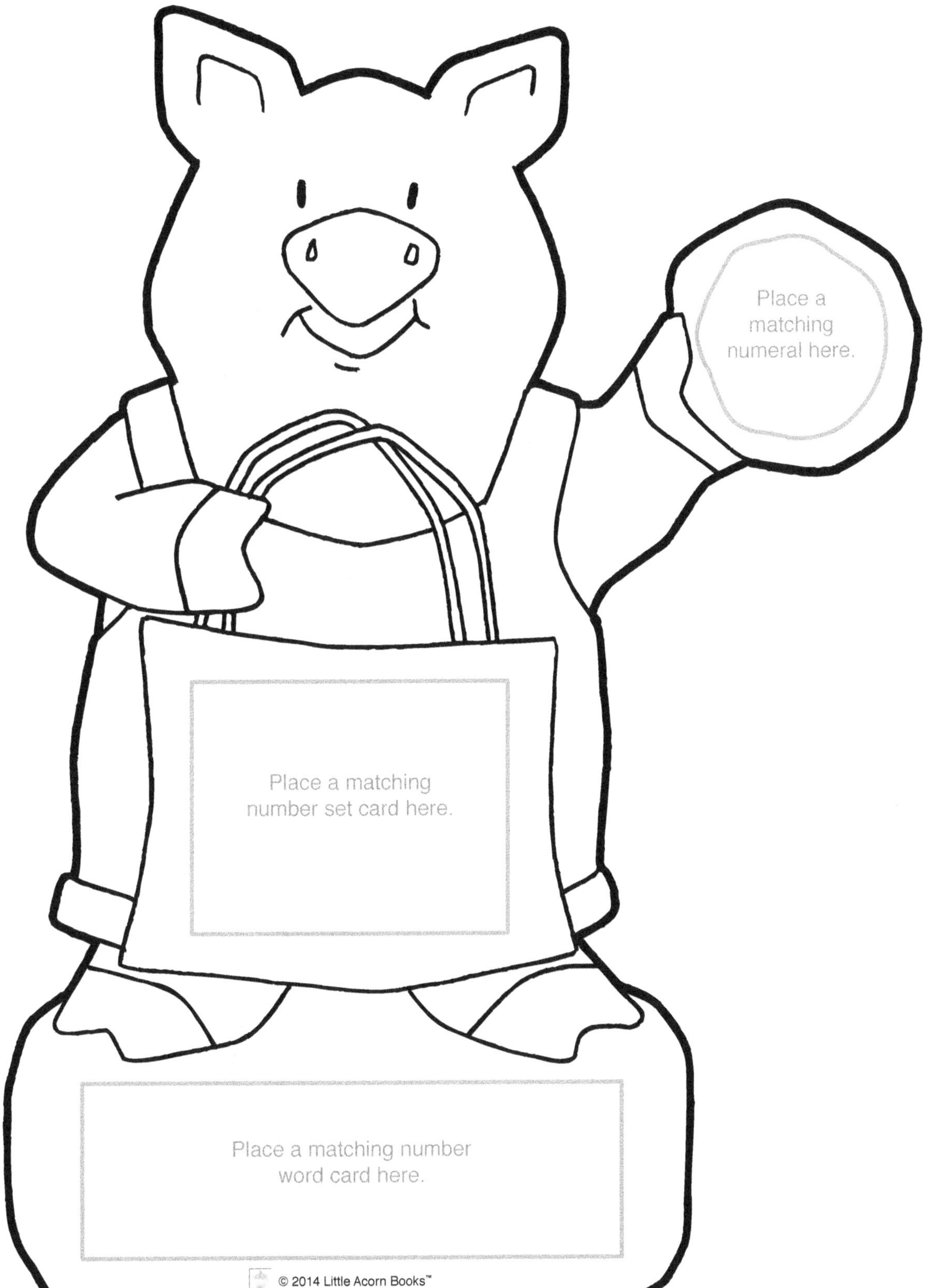
Place a matching numeral here.
Place a matching number set card here.
Place a matching number word card here.
© 2014 Little Acorn Books™

Koala Match Board

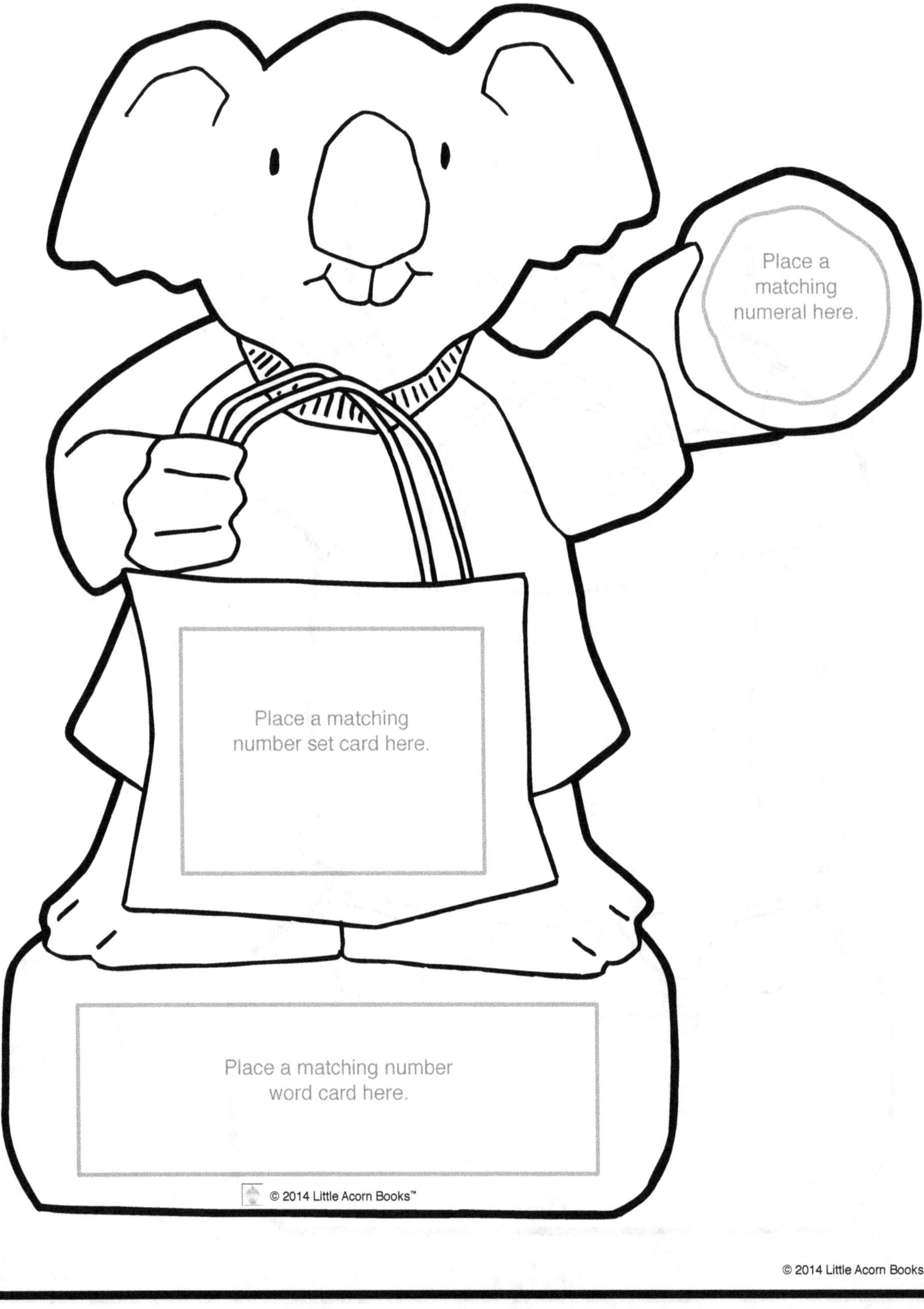

Place a matching numeral here.

Place a matching number set card here.

Place a matching number word card here.

Snake Match Board

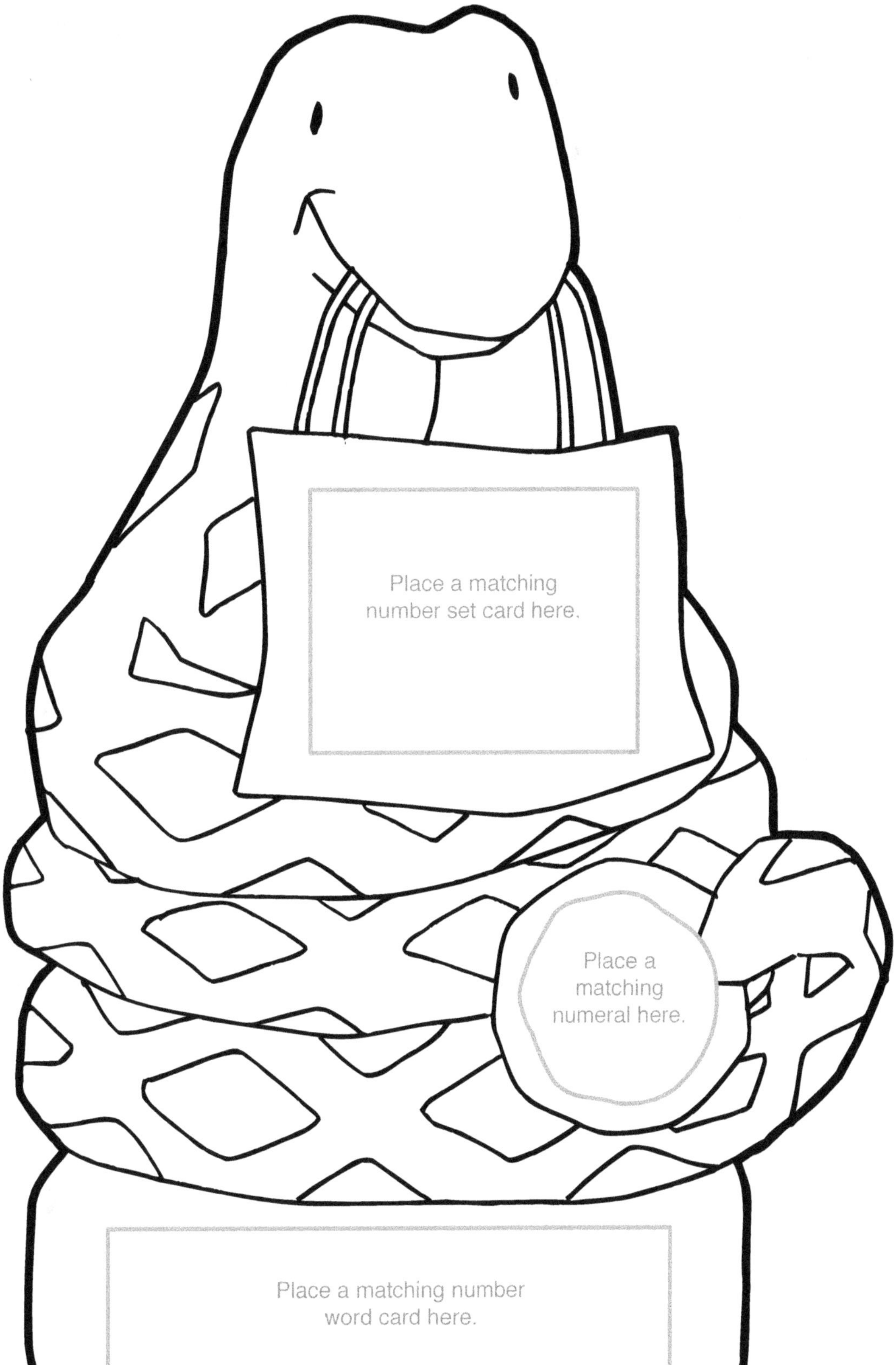
Place a matching
number set card here.
Place a
matching
numeral here.
Place a matching number
word card here.
© 2014 Little Acorn Books™

Butterfly Match Board

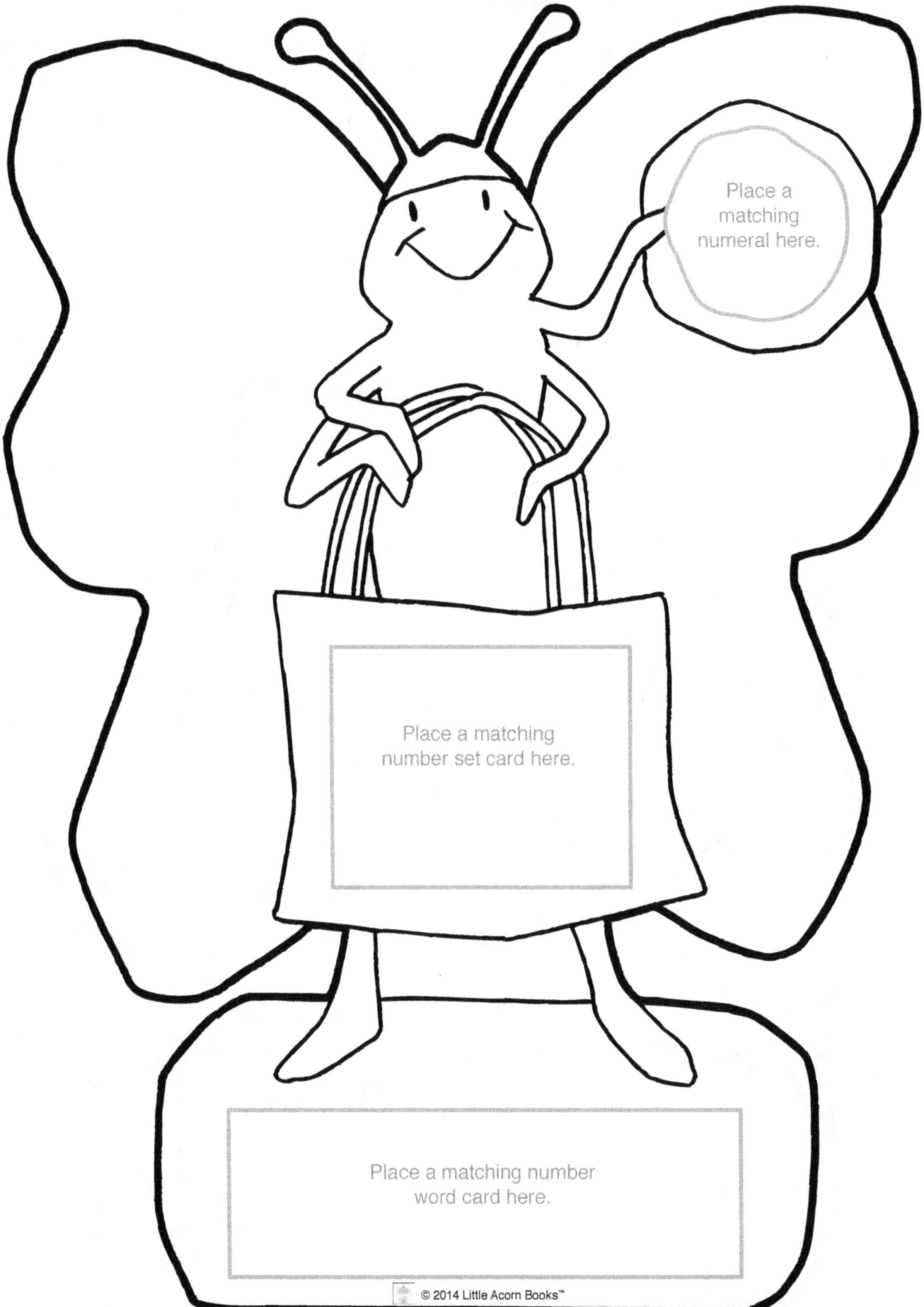

Place a matching numeral here.
Place a matching number set card here.
Place a matching number word card here.
© 2014 Little Acorn Books™

Rabbit Match Board

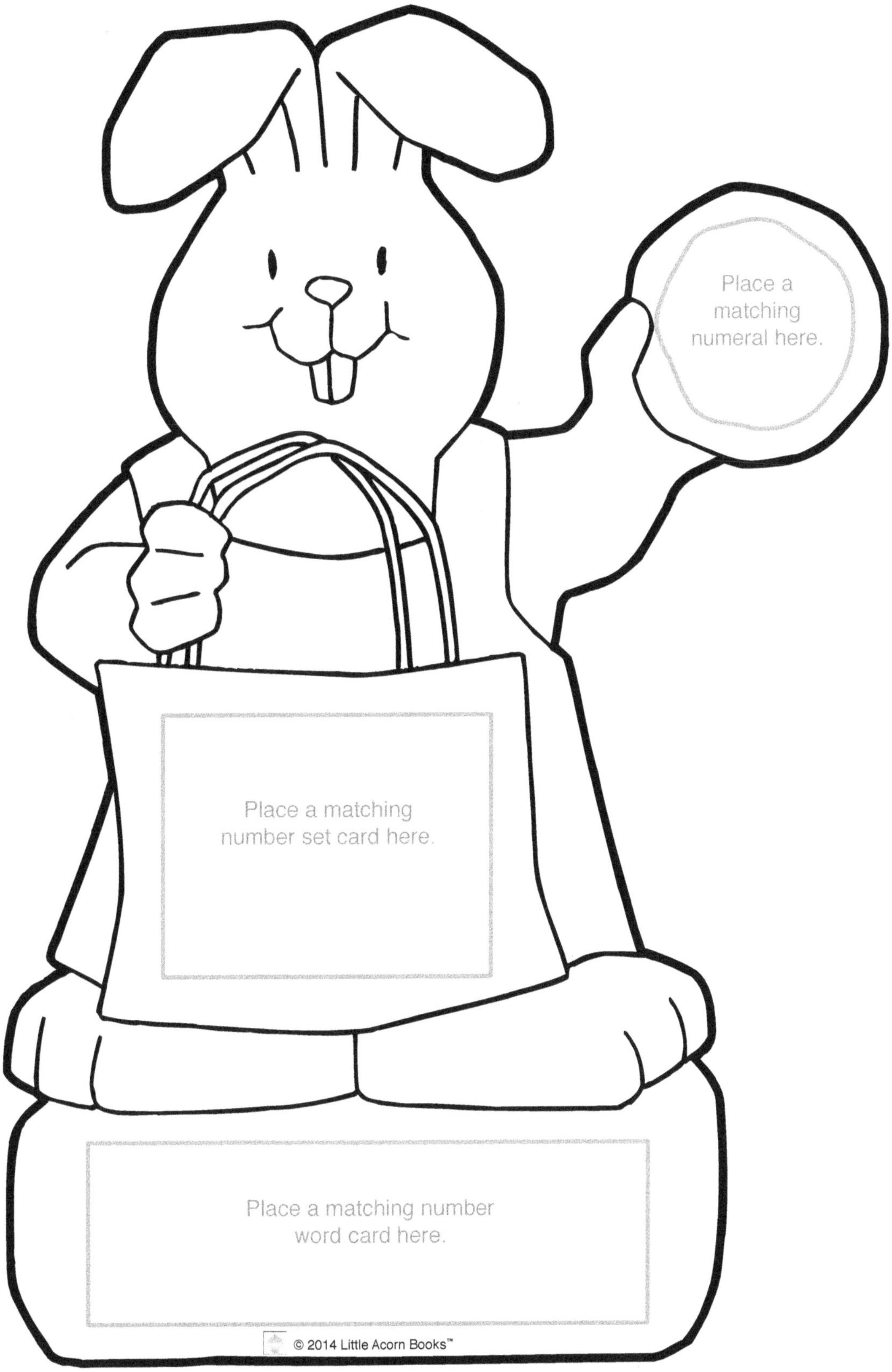
Place a matching numeral here.
Place a matching number set card here.
Place a matching number word card here.
© 2014 Little Acorn Books™

Frog Match Board

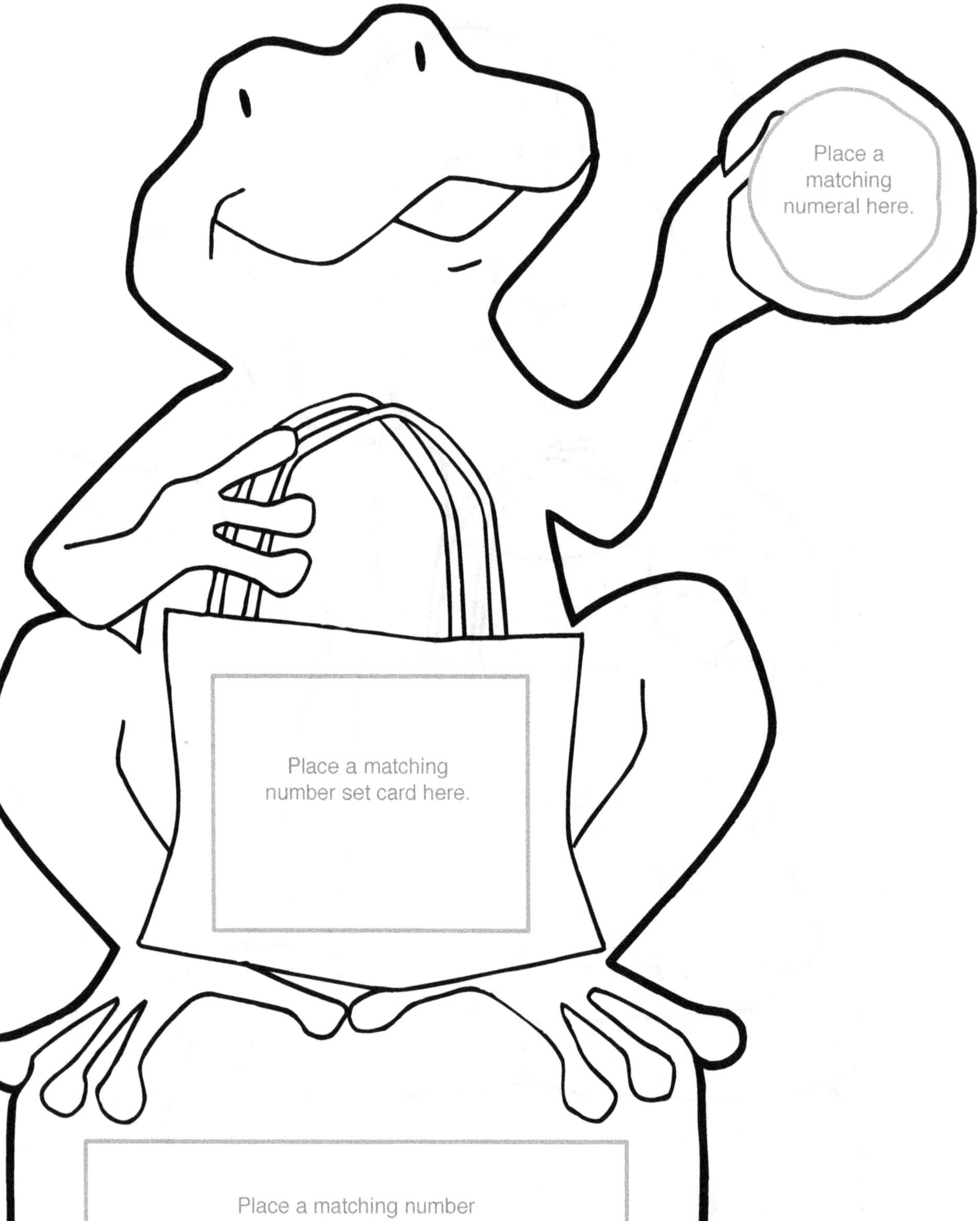
Place a matching numeral here.
Place a matching number set card here.
Place a matching number word card here.
© 2014 Little Acorn Books™

Turtle Match Board

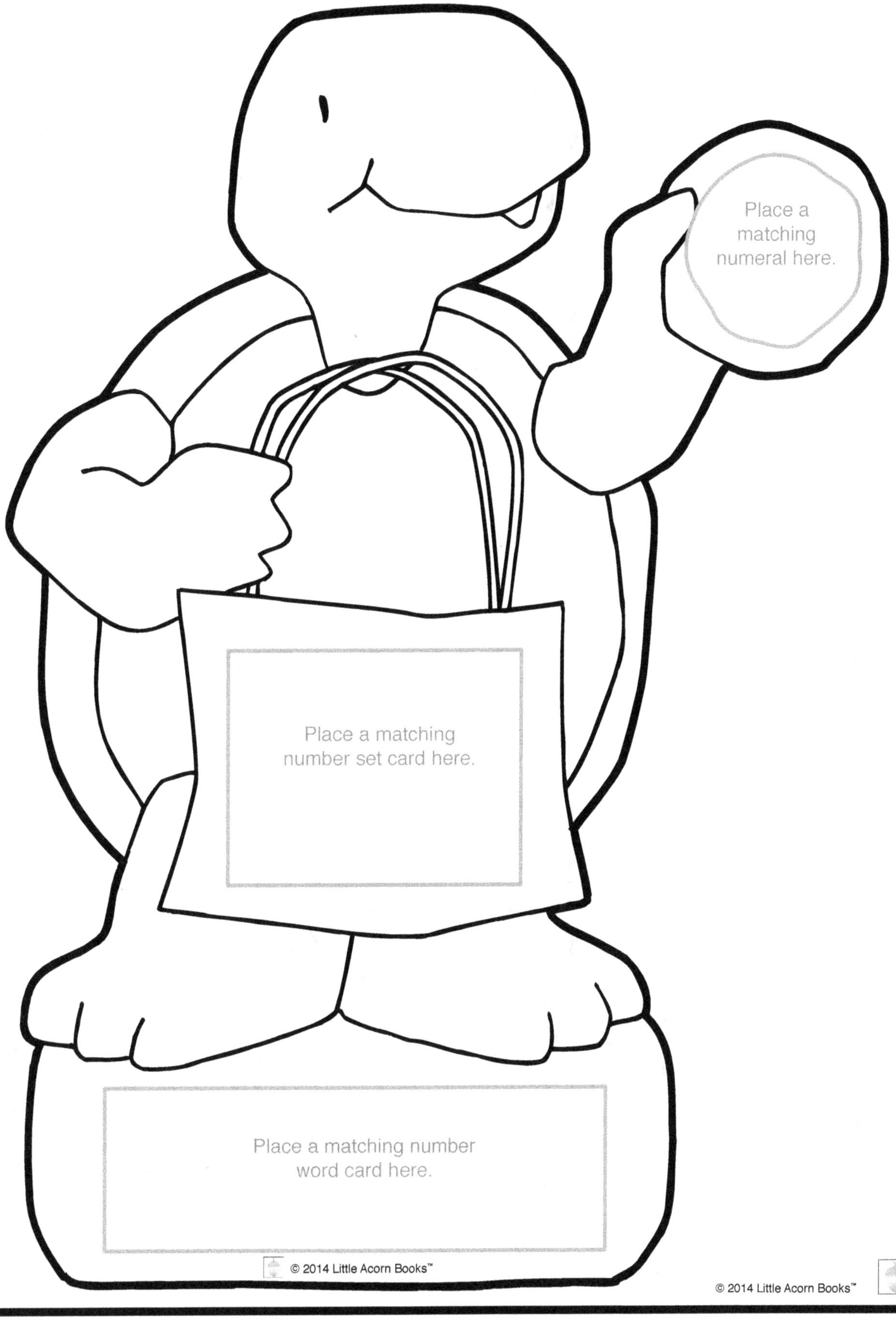
Place a matching numeral here.
Place a matching number set card here.
Place a matching number word card here.
© 2014 Little Acorn Books™

Peanut Cards

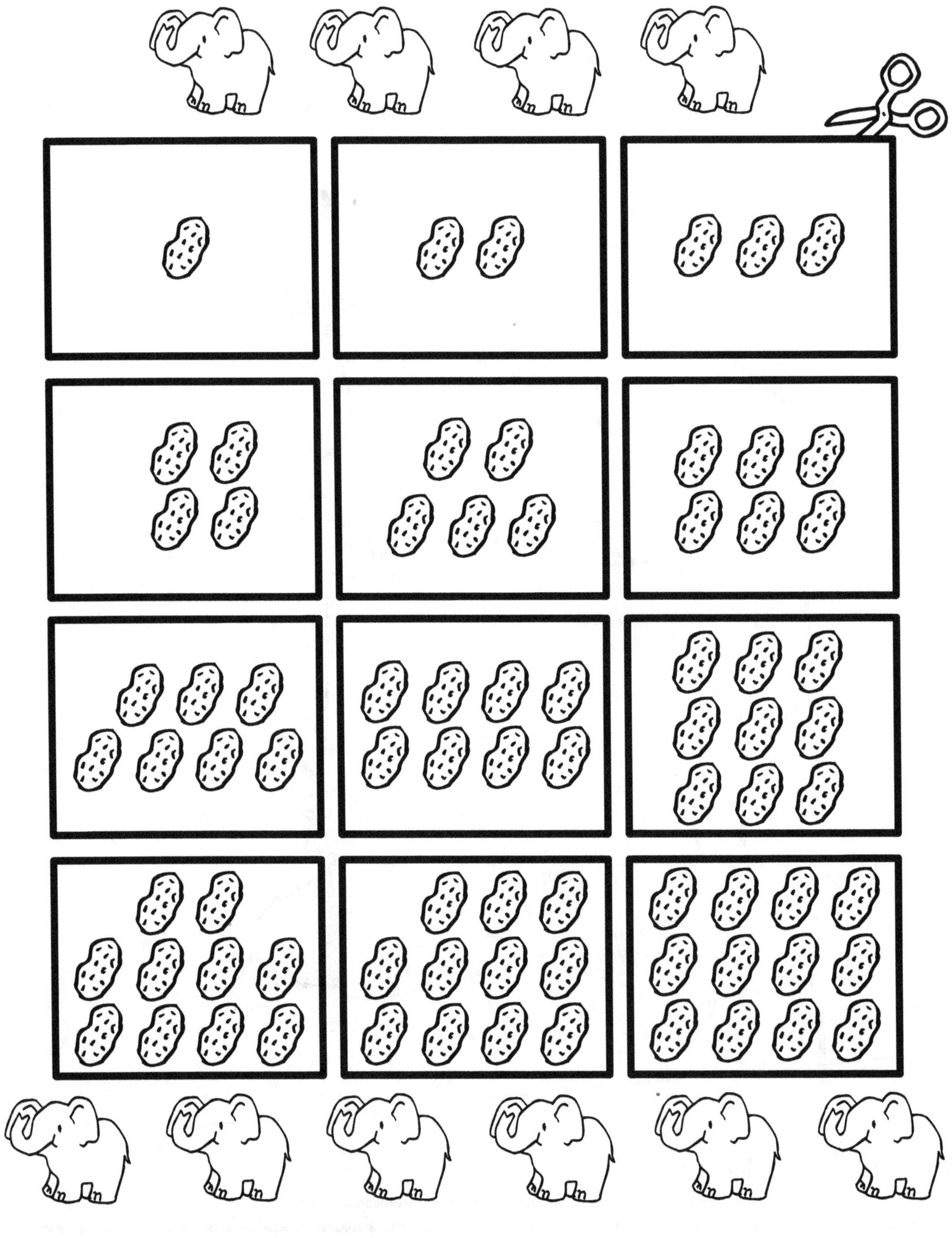

Cookie Cards

Button Cards

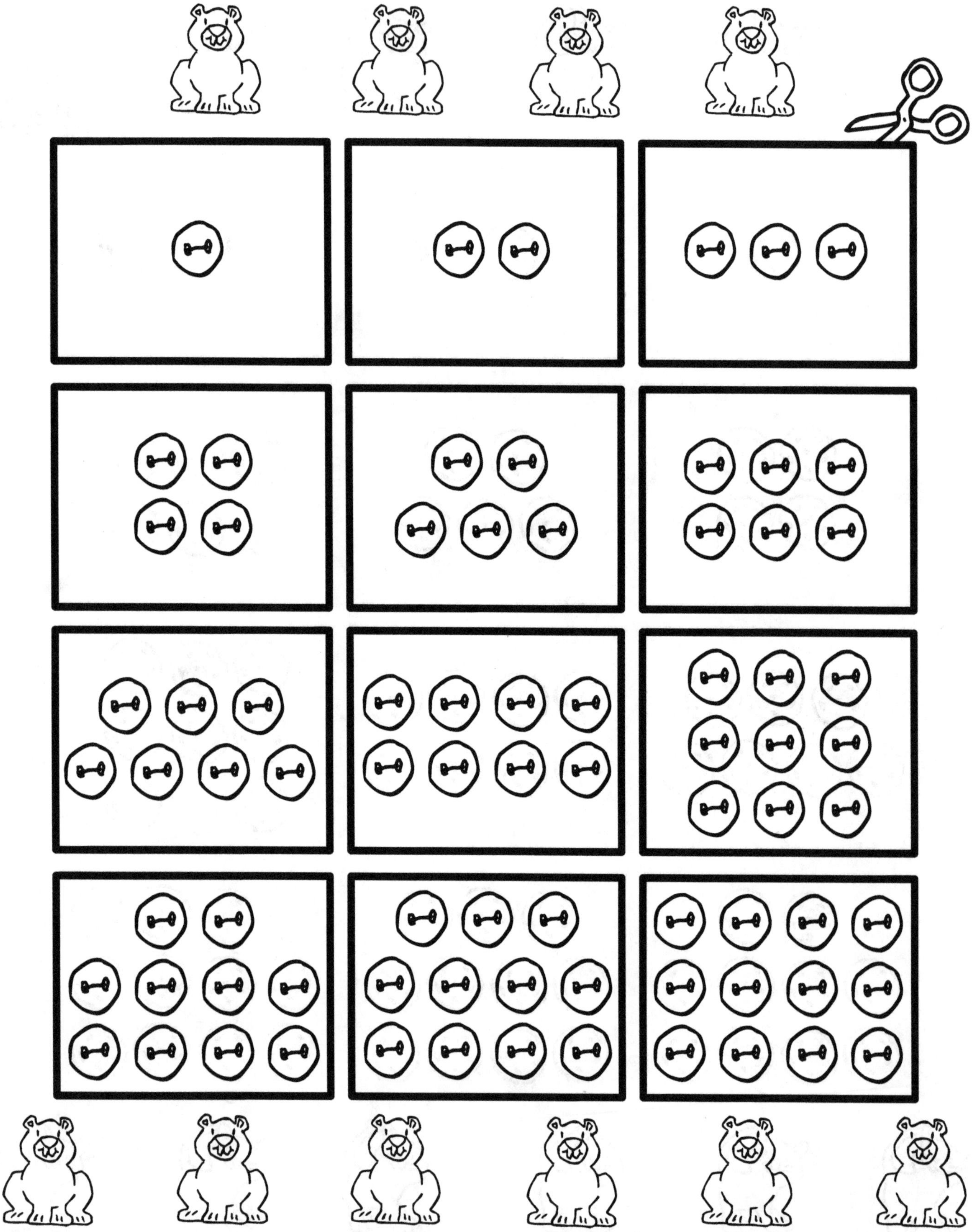

Drum Cards

Bell Cards

34

Flower Cards

LAB201410P • About Early Math • 978-1-937257-56-9 • © 2014 Little Acorn Books™

Key Cards

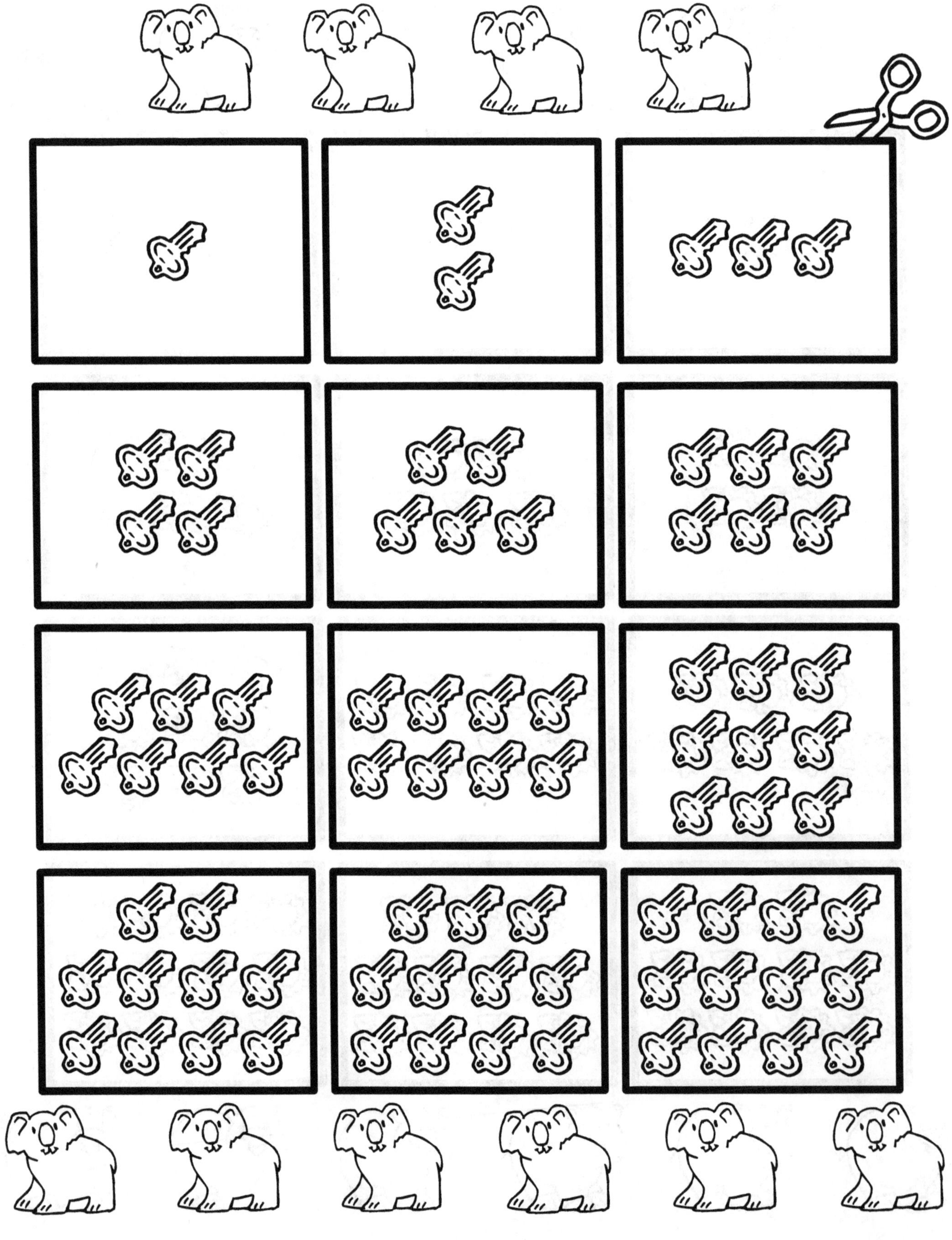

Star Cards

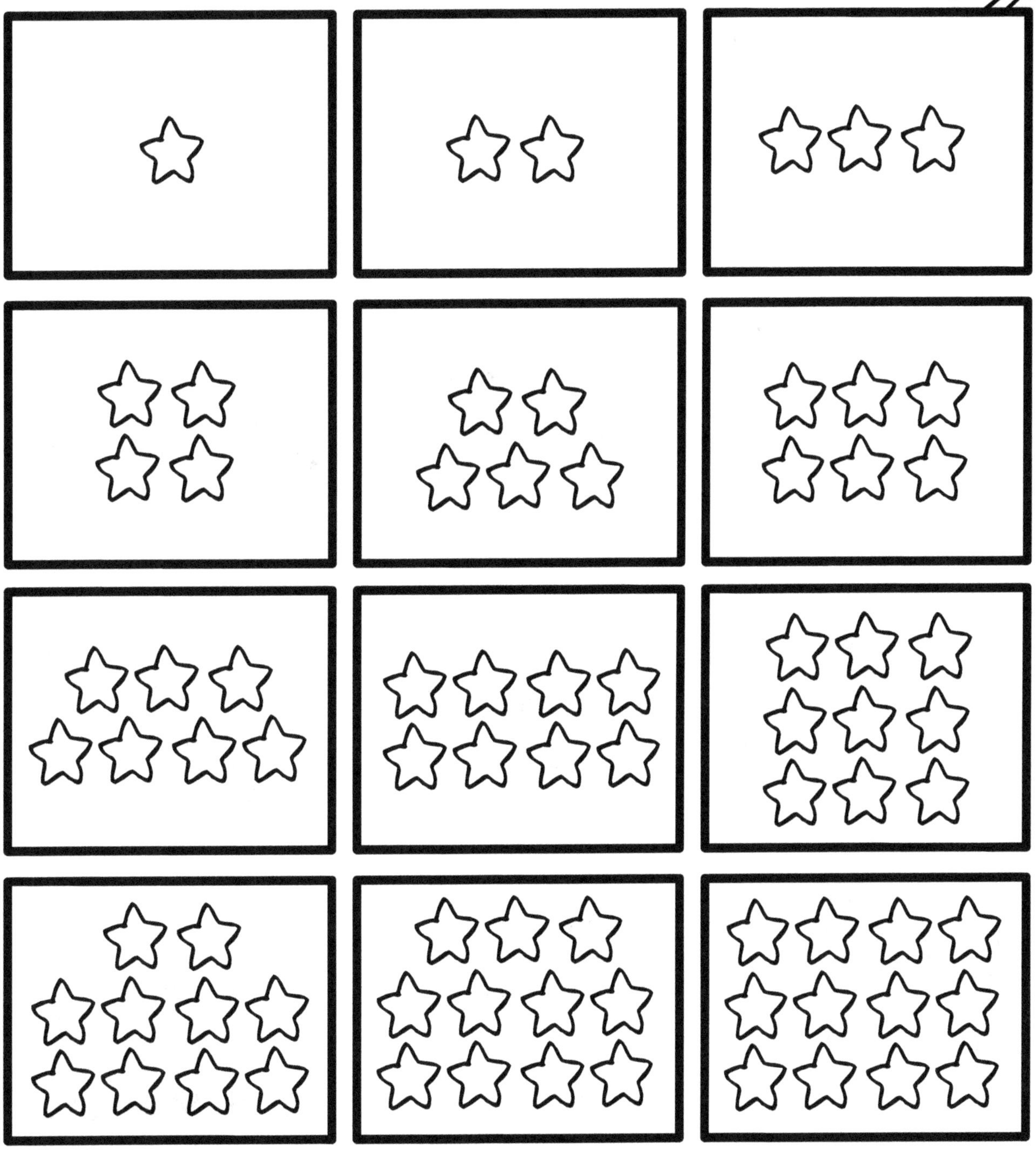

Bow Cards

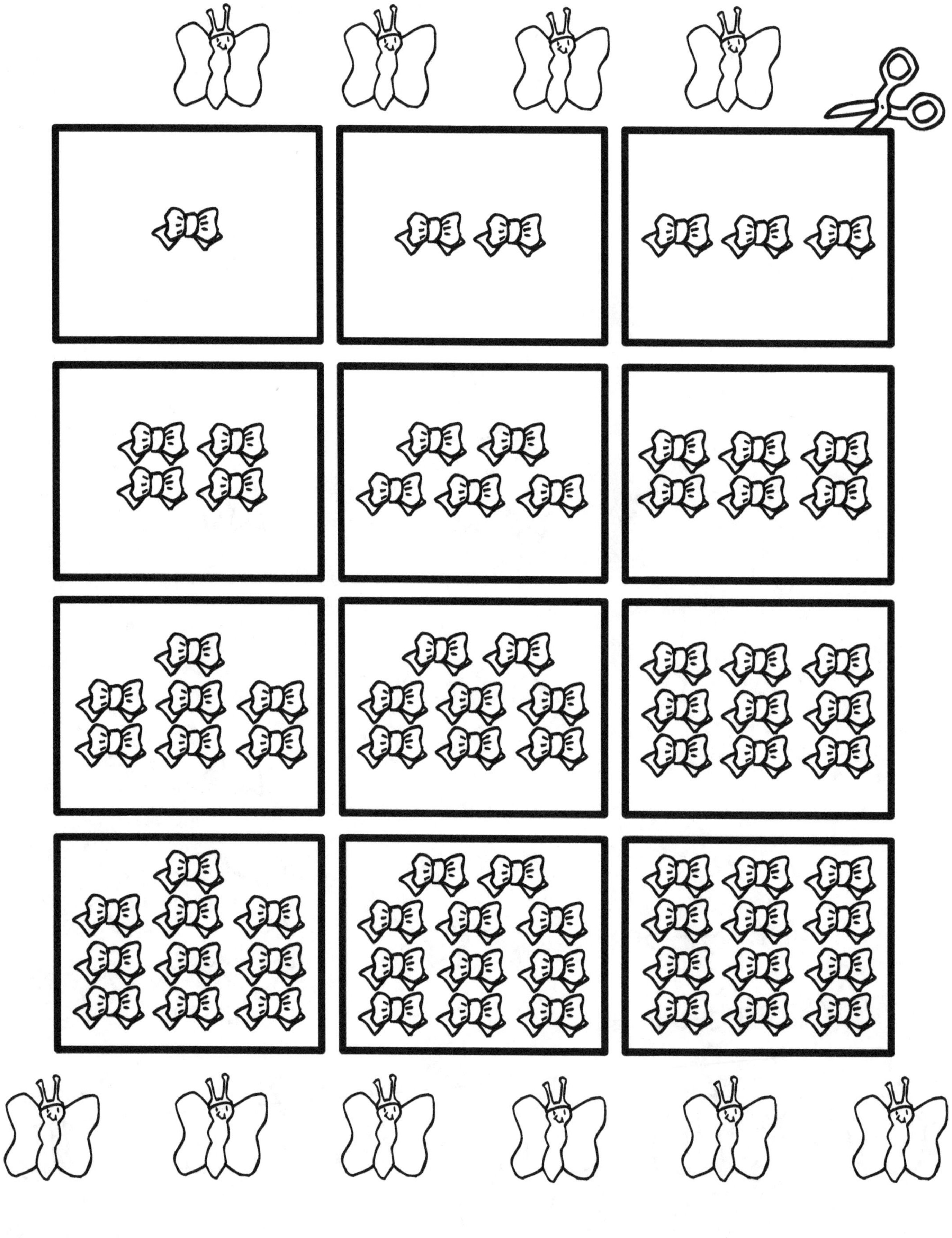

Carrot Cards

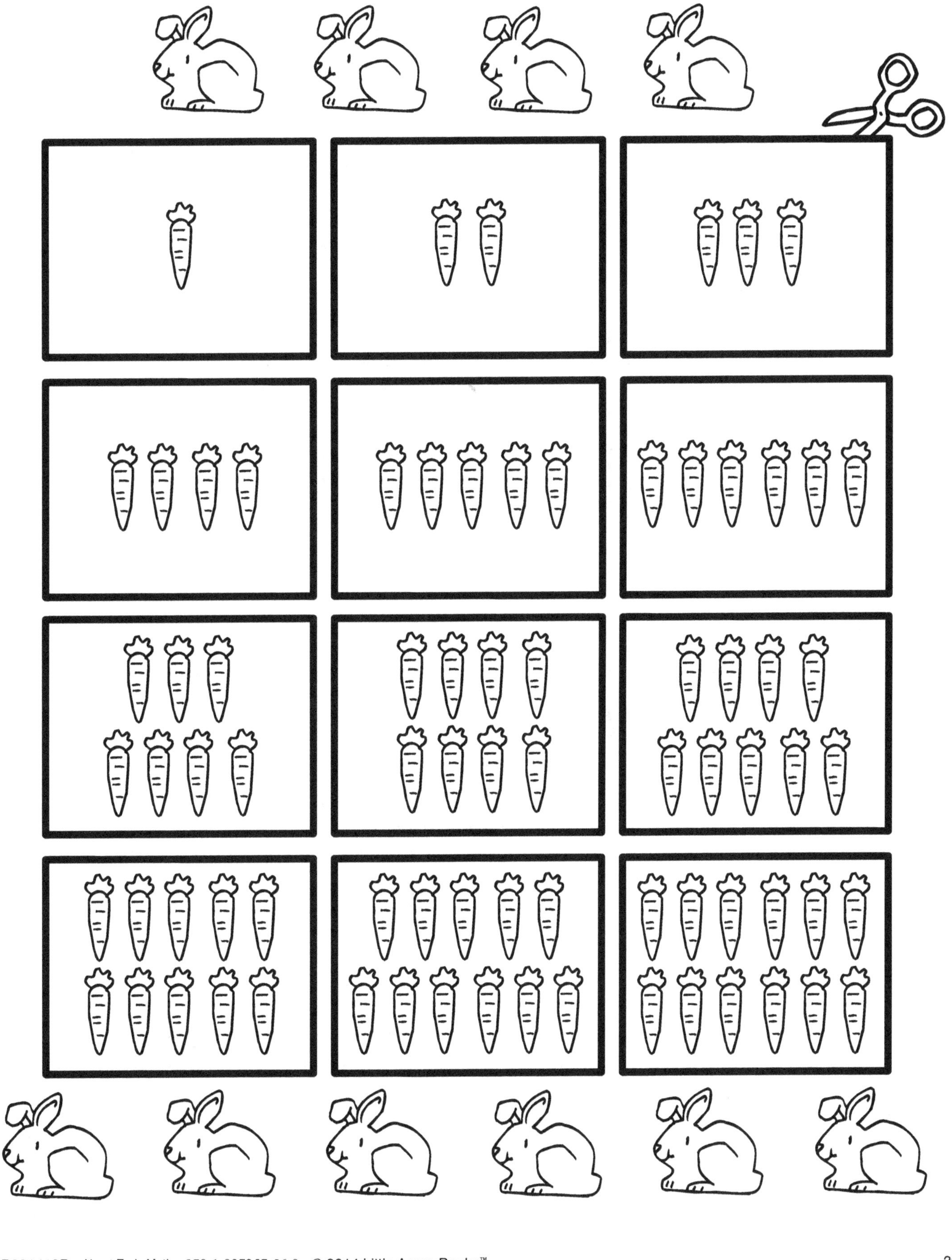

Balloon Cards

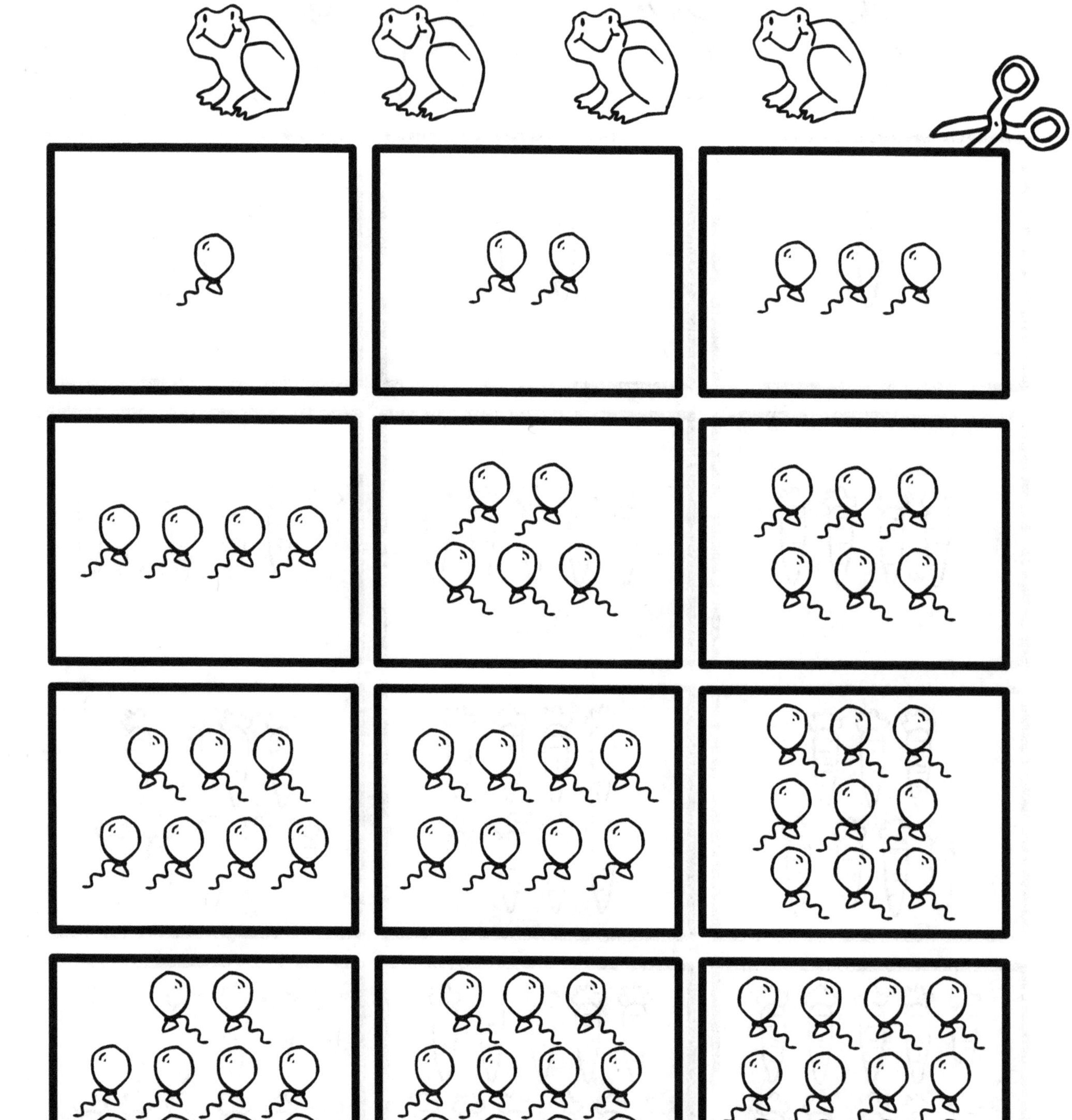

Heart Cards

Numeral Cards

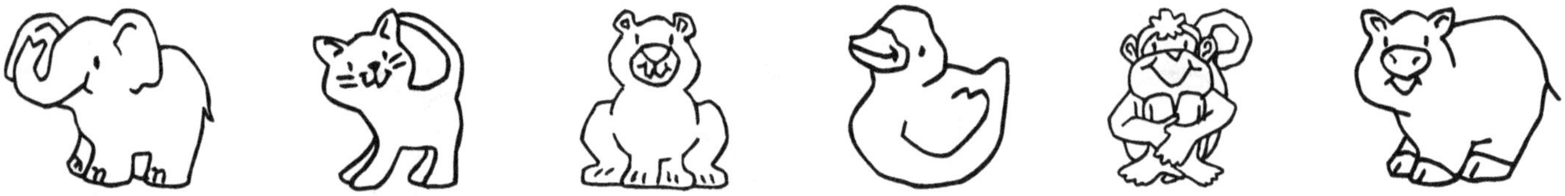

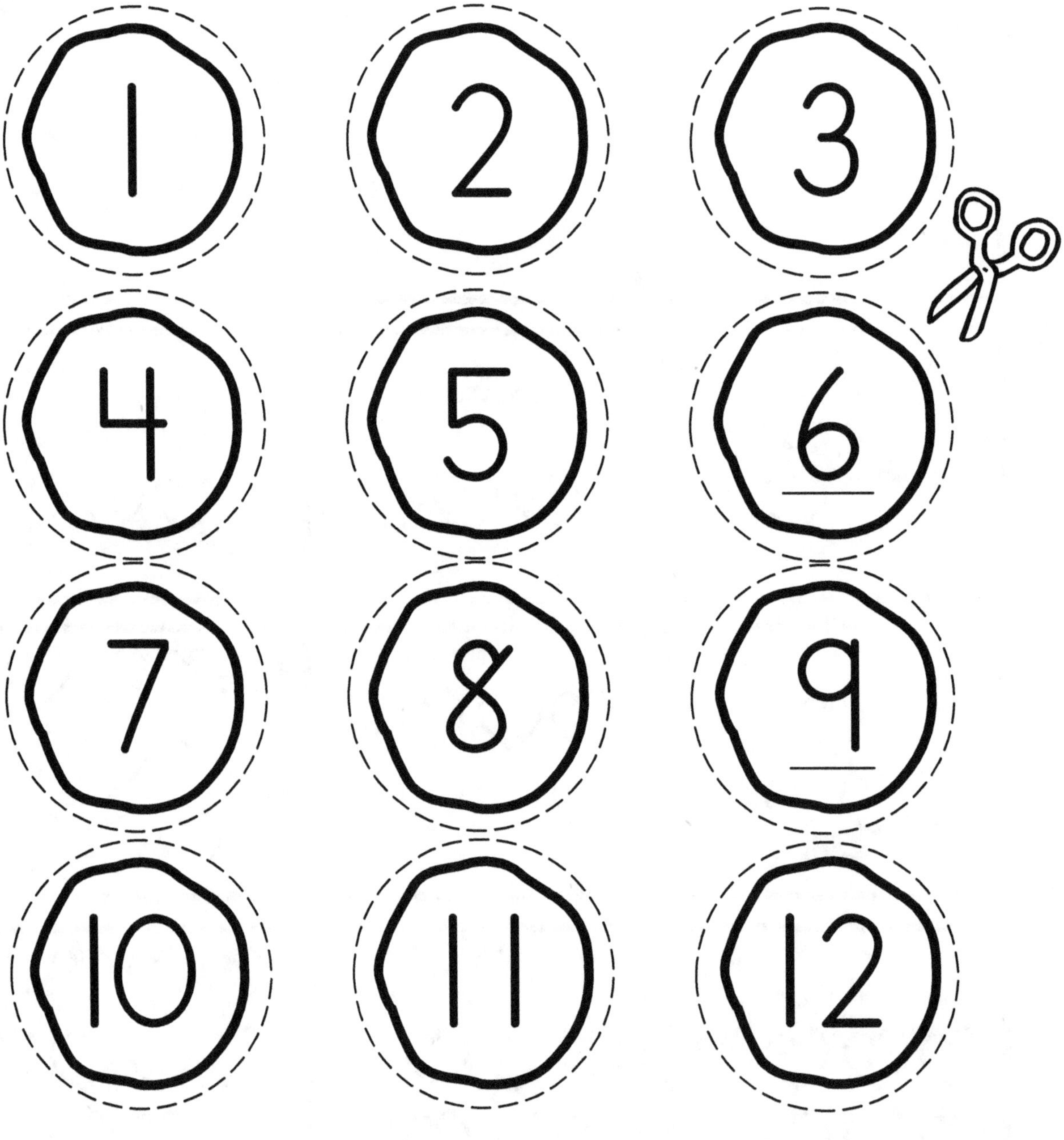

1 2 3
4 5 6
7 8 9
10 11 12

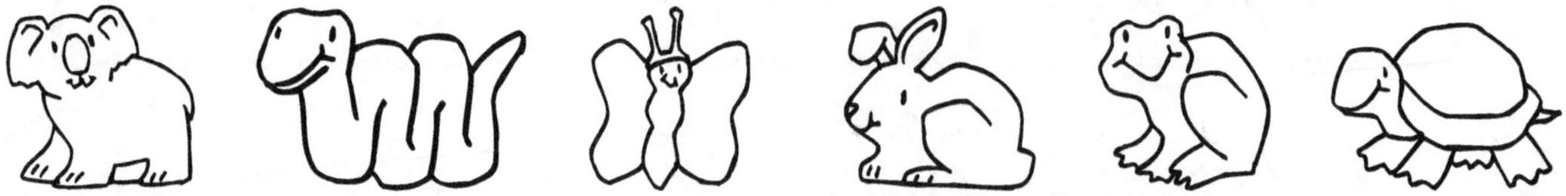

Number Word Cards

one	two
three	four
five	six
seven	eight
nine	ten
eleven	twelve

Add or Take Away

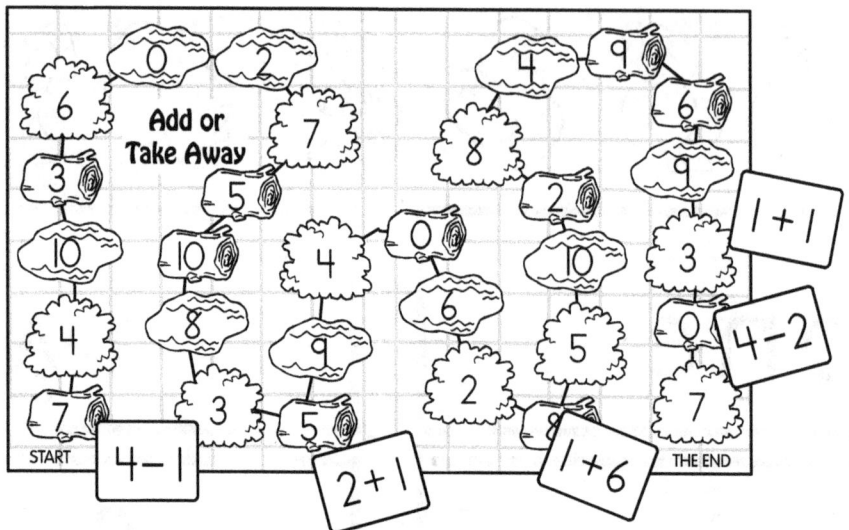

Reproduce, color, and cut out the game board patterns, pawns, and cards. Matching in the center, glue the game board patterns to the inside of a folder. Laminate the game board folder, pawns, and game cards. Cut apart the pawns and game cards. Decorate the cover and tape an envelope to the back of the folder to store pawns and game cards. Option: Reproduce, color, and glue each page of cards to the back of a sheet of gift wrap, then laminate, and cut apart the cards.

To play, each of two to four players chooses one of the twelve animal pawns. One player shuffles and places the deck of cards, face down, on the table. Each player, in turn, draws a card, solves the addition or subtraction problem on the card, and moves his or her pawn to the next matching numeral space on the board. Drawn cards are placed, face down, in a discard pile. Play continues until each player reaches The End. When all the cards have been drawn, reshuffle the discard pile and continue playing.
Option: Separate the cards into addition only, subtraction only.

Pawns

Add or Take Away Game Board

Add or Take Away

START

Add or Take Away Game Board

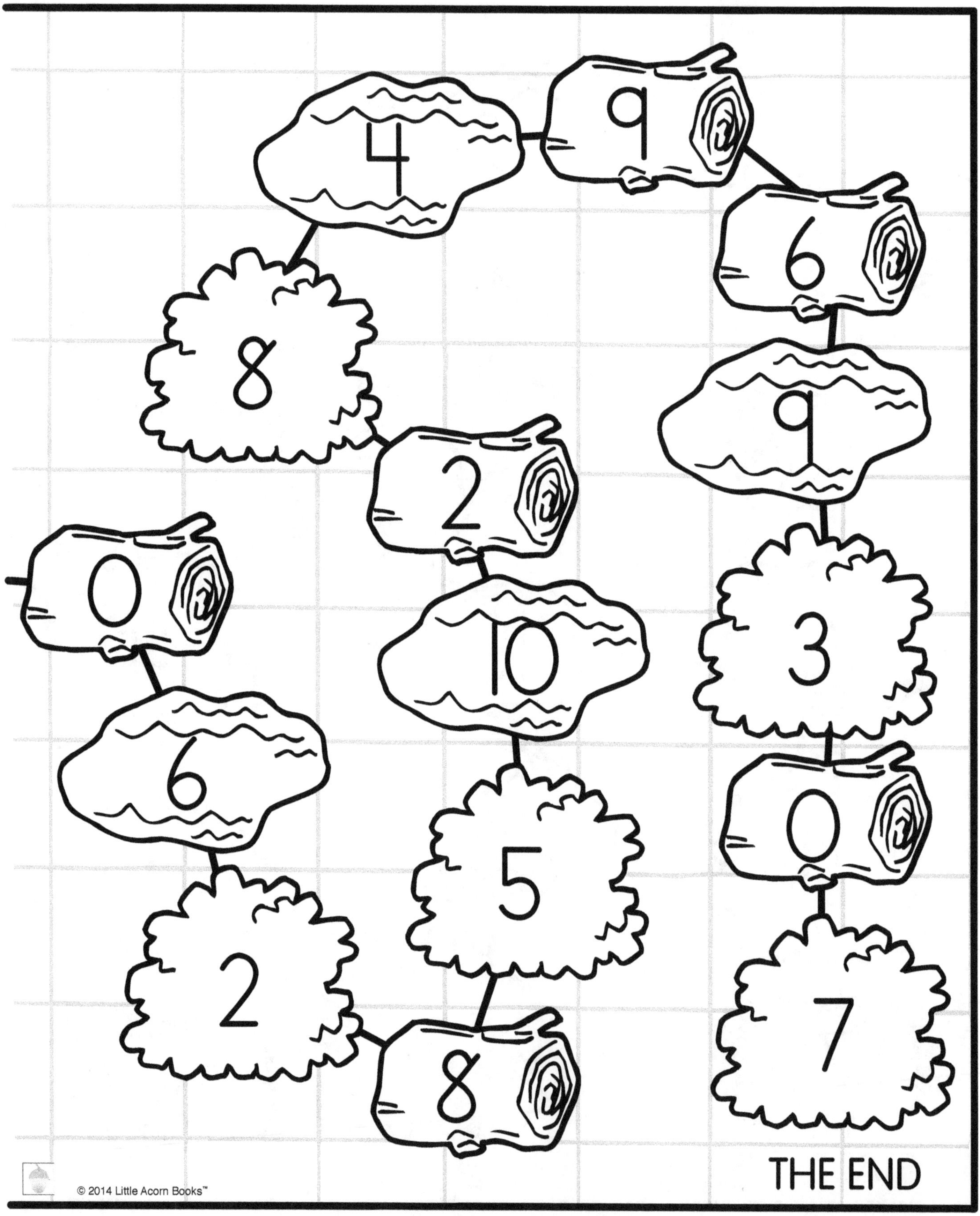

THE END

Addition Problem Cards

1+0	1+1	1+2
1+3	1+4	1+5
1+6	1+7	1+8
1+9	2+0	2+1
2+2	2+3	2+4
2+5	2+6	2+7

Addition Problem Cards

2+8	3+0	3+1
3+2	3+3	3+4
3+5	3+6	3+7
4+0	4+1	4+2
4+3	4+4	4+5
4+6	5+0	5+1

Addition Problem Cards

5+2	5+3	5+4
5+5	6+0	6+1
6+2	6+3	6+4
7+0	7+1	7+2
7+3	8+0	8+1
8+2	9+0	9+1

Subtraction Problem Cards

9 − 9	9 − 8	9 − 7
9 − 6	9 − 5	9 − 4
9 − 3	9 − 2	9 − 1
9 − 0	8 − 8	8 − 7
8 − 6	8 − 5	8 − 4
8 − 3	8 − 2	8 − 1

Subtraction Problem Cards

8 − 0	7 − 7	7 − 6
7 − 5	7 − 4	7 − 3
7 − 2	7 − 1	7 − 0
6 − 6	6 − 5	6 − 4
6 − 3	6 − 2	6 − 1
6 − 0	5 − 5	5 − 4

Subtraction Problem Cards

5−3	5−2	5−1
5−0	4−4	4−3
4−2	4−1	4−0
3−3	3−2	3−1
3−0	2−2	2−1
2−0	1−1	1−0

Math Problem Boards

Number of the Day Badges

The Number of the Day

The Number of the Day

The Number of the Day

The Number of the Day

Numbers Chart

1	2	3	4
5	6	7	8
9	10	11	12

Number Words Chart

one	two	three	four
five	six	seven	eight
nine	ten	eleven	twelve

Awards Booklet

My Math Awards Booklet

I can write numbers.

1 2 3 4 5
6 7 8 9

Glue a star sticker here.

I can count to _____.
numeral

Glue a star sticker here.

I can count number sets.

Glue a star sticker here.

I can color 10 things.

Glue a star sticker here.

I can read number words.

one   two
three

Glue a star sticker here.

Awards Booklet

I can match numbers and number sets.

3 •••

Glue a star sticker here.

I can match numbers and number words.

4 four

Glue a star sticker here.

I can add.

🐘 + 🐘 =

Glue a star sticker here.

I can take away.

🐱🐱 − 🐱 =

Glue a star sticker here.

Awards

Name

can count to

100!

© 2014 Little Acorn Books™

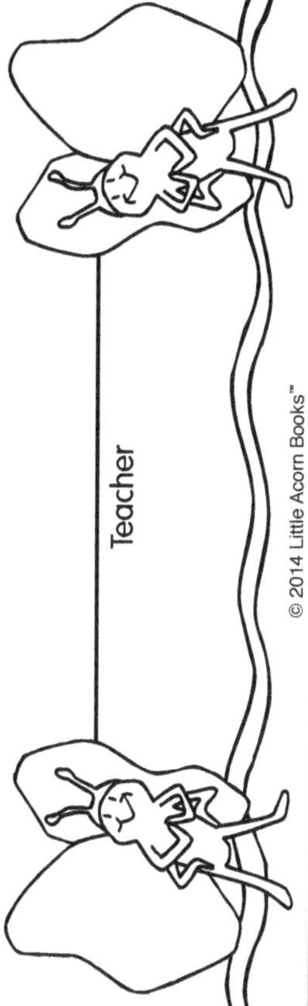

Name

Good Work

Early Math Skills Achievement

Teacher

© 2014 Little Acorn Books™

Awarded to

Name

for

Super Math Skills

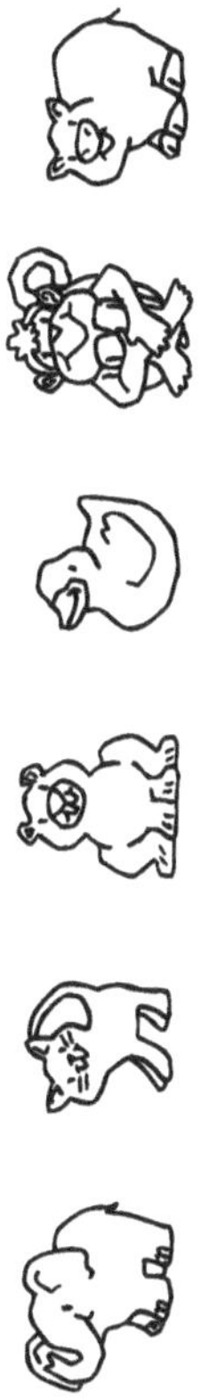

© 2014 Little Acorn Books™

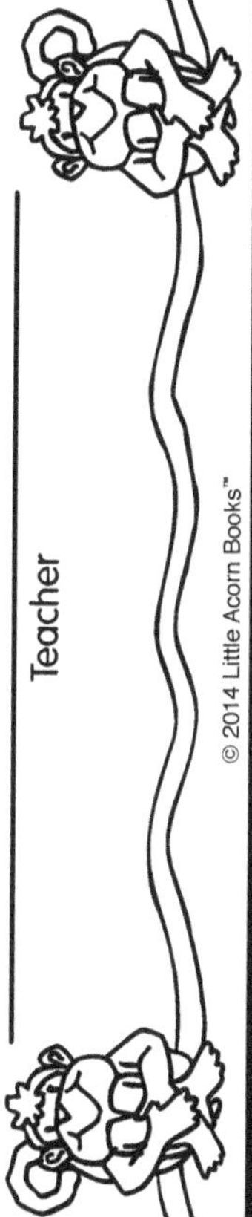

Name

is awarded this certificate for

Marvelous

Math Skills Achievement

Teacher

© 2014 Little Acorn Books™

Take-Home Notes

Clock Form

Reproduce, color, cut out, and laminate the clock and hands. Punch a hole in the center of the clock and at the round end of each hand. Attach the hands to the clock with a brass fastener.

Counters

1 2 3 4 5 6 7 8 9 0 1 2 3 4 5 6 7 8 9 0

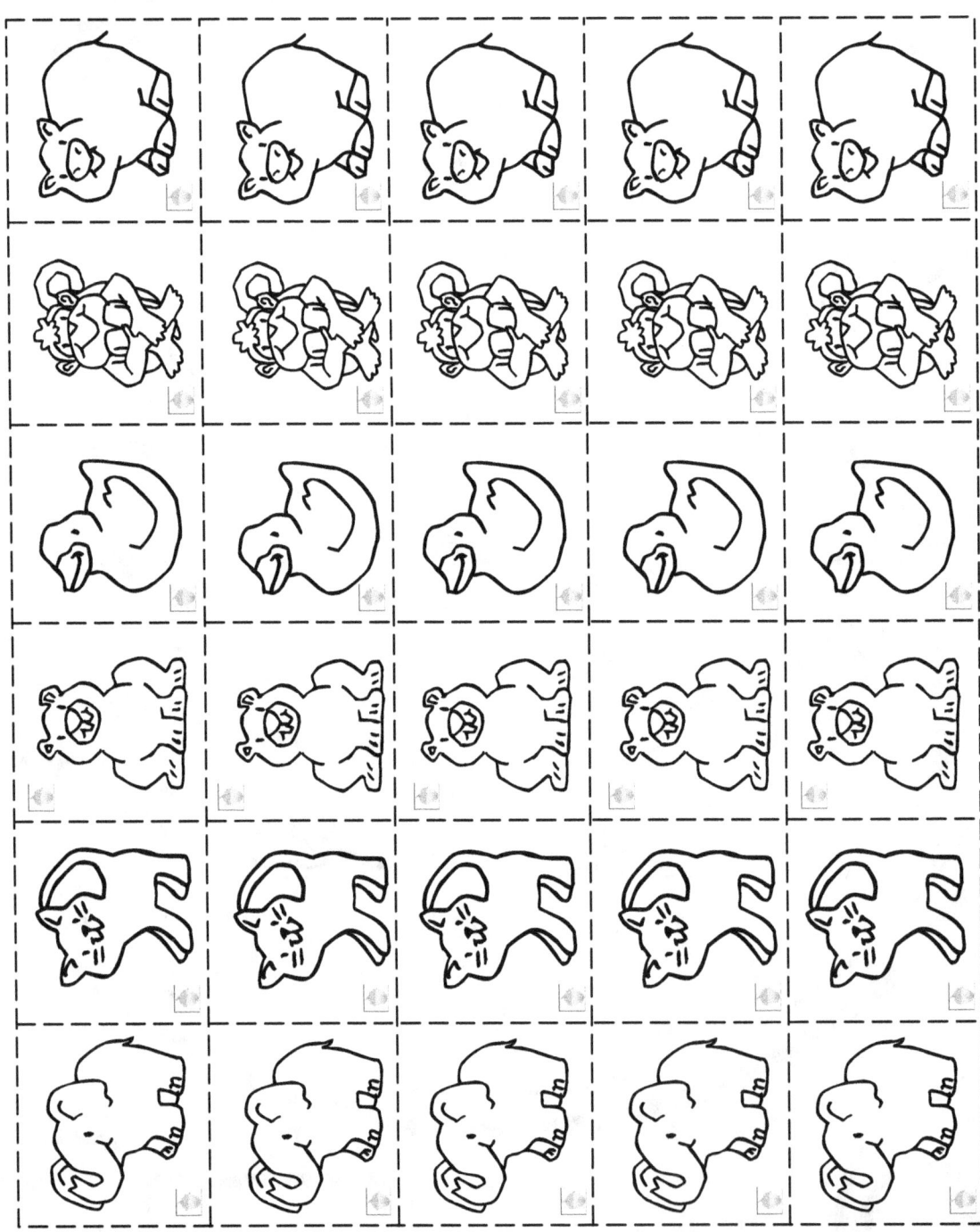

1 2 3 4 5 6 7 8 9 0 1 2 3 4 5 6 7 8 9 0

Reproduce, laminate, and cut apart 4 sets of construction paper counters.

Counters

1 2 3 4 5 6 7 8 9 0 1 2 3 4 5 6 7 8 9 0

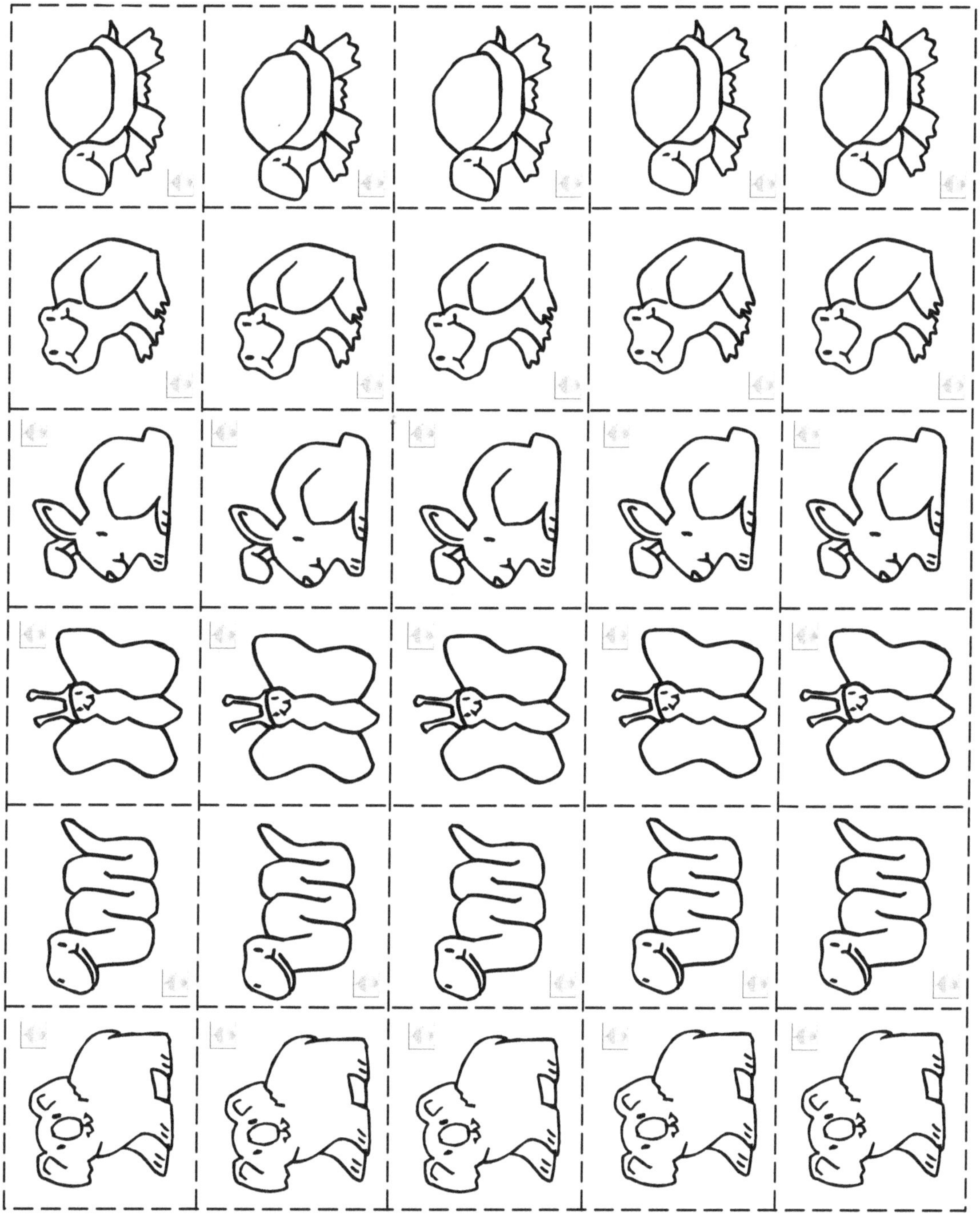

1 2 3 4 5 6 7 8 9 0 1 2 3 4 5 6 7 8 9 0

Reproduce, laminate, and cut apart 4 sets of construction paper counters.

Animal Inches

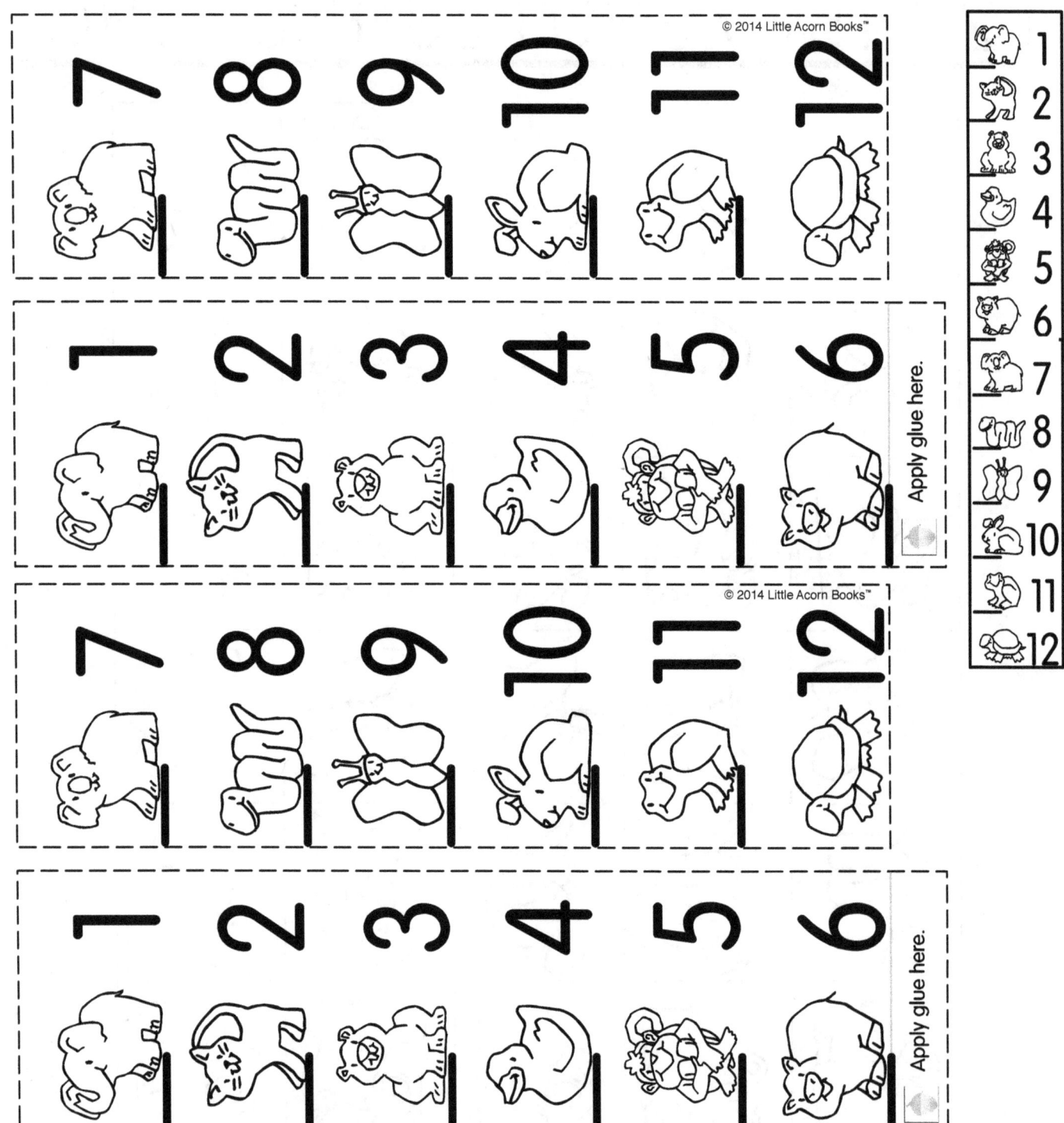

Reproduce, color, cut out, and assemble to form two 12-inch rulers.
Teach children how to measure familiar objects.

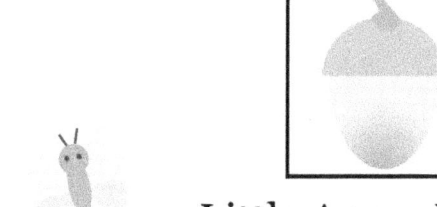

Little Acorn Books™

Promoting Early Skills for a Lifetime™

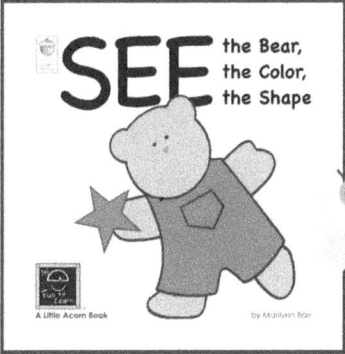 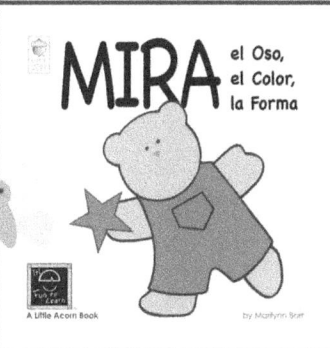  

A Hands-on Picture Book Series • Infancy–Age 4

Miss Pitty Pat & Friends
Preschool–Grade 1

Mookie's Christmas Tree
For All Ages and
Not Just for Christmas

Using Crayons, Scissors, & Glue for Crafts
Preschool–Grade 1

Little Acorn Books™
Visit our web site:
www.littleacornbooks.com

www.ingramcontent.com/pod-product-compliance
Lightning Source LLC
Chambersburg PA
CBHW081455060426
42444CB00037BA/3289